Dynamic Speech Models, Theory, Algorithms, and Applications
Li Deng

978-3-031-01427-7 paper Deng
978-3-031-02555-6 ebook Deng

DOI: 10.1007/978-3-031-02555-6

A Publication in the Springer series
SYNTHESIS LECTURES ON SPEECH AND AUDIO PROCESSING
Lecture #2
Series editor B. H. Juang

First Edition
10 9 8 7 6 5 4 3 2 1

Dynamic Speech Models
Theory, Algorithms, and Applications

Li Deng
Microsoft Research
Redmond, Washington, USA

SYNTHESIS LECTURES ON SPEECH AND AUDIO PROCESSING #2

ABSTRACT

Speech dynamics refer to the temporal characteristics in all stages of the human speech communication process. This speech "chain" starts with the formation of a linguistic message in a speaker's brain and ends with the arrival of the message in a listener's brain. Given the intricacy of the dynamic speech process and its fundamental importance in human communication, this monograph is intended to provide a comprehensive material on mathematical models of speech dynamics and to address the following issues: How do we make sense of the complex speech process in terms of its functional role of speech communication? How do we quantify the special role of speech timing? How do the dynamics relate to the variability of speech that has often been said to seriously hamper automatic speech recognition? How do we put the dynamic process of speech into a quantitative form to enable detailed analyses? And finally, how can we incorporate the knowledge of speech dynamics into computerized speech analysis and recognition algorithms? The answers to all these questions require building and applying computational models for the dynamic speech process.

What are the compelling reasons for carrying out dynamic speech modeling? We provide the answer in two related aspects. First, scientific inquiry into the human speech code has been relentlessly pursued for several decades. As an essential carrier of human intelligence and knowledge, speech is the most natural form of human communication. Embedded in the speech code are linguistic (as well as para-linguistic) messages, which are conveyed through four levels of the speech chain. Underlying the robust encoding and transmission of the linguistic messages are the speech dynamics at all the four levels. Mathematical modeling of speech dynamics provides an effective tool in the scientific methods of studying the speech chain. Such scientific studies help understand why humans speak as they do and how humans exploit redundancy and variability by way of multitiered dynamic processes to enhance the efficiency and effectiveness of human speech communication. Second, advancement of human language technology, especially that in automatic recognition of natural-style human speech is also expected to benefit from comprehensive computational modeling of speech dynamics. The limitations of current speech recognition technology are serious and are well known. A commonly acknowledged and frequently discussed weakness of the statistical model underlying current speech recognition technology is the lack of adequate dynamic modeling schemes to provide correlation structure across the temporal speech observation sequence. Unfortunately, due to a variety of reasons, the majority of current research activities in this area favor only incremental modifications and improvements to the existing HMM-based state-of-the-art. For example, while the dynamic and correlation modeling is known to be an important topic, most of the systems nevertheless employ only an ultra-weak form of speech dynamics; e.g., differential or delta parameters. Strong-form dynamic speech modeling, which is the focus of this monograph, may serve as an ultimate solution to this problem.

After the introduction chapter, the main body of this monograph consists of four chapters. They cover various aspects of theory, algorithms, and applications of dynamic speech models, and provide a comprehensive survey of the research work in this area spanning over past 20 years. This monograph is intended as advanced materials of speech and signal processing for graudate-level teaching, for professionals and engineering practioners, as well as for seasoned researchers and engineers specialized in speech processing.

KEYWORDS

Articulatory trajectories, Automatic speech recognition, Coarticulation, Discretizing hidden dynamics, Dynamic Bayesian network, Formant tracking, Generative modeling, Speech acoustics, Speech dynamics, Vocal tract resonance

Contents

Acknowledgments

This book would not have been possible without the help and support from friends, family, colleagues, and students. Some of the material in this book is the result of collaborations with my former students and current colleagues. Special thanks go to Jeff Ma, Leo Lee, Dong Yu, Alex Acero, Jian-Lai Zhou, and Frank Seide.

The most important acknowledgments go to my family. I also thank Microsoft Research for providing the environment in which the research described in this book is made possible. Finally, I thank Prof. Fred Juang and Joel Claypool for not only the initiation but also the encouragement and help throughout the course of writting this book.

CHAPTER 1

Introduction

1.1 WHAT ARE SPEECH DYNAMICS?

In a broad sense, speech dynamics are time-varying or temporal characteristics in all stages of the human speech communication process. This process, sometimes referred to as speech chain [1], starts with the formation of a linguistic message in the speaker's brain and ends with the arrival of the message in the listener's brain. In parallel with this direct information transfer, there is also a feedback link from the acoustic signal of speech to the speaker's ear and brain. In the conversational mode of speech communication, the style of the speaker's speech can be further influenced by an assessment of the extent to which the linguistic message is successfully transferred to or understood by the listener. This type of feedbacks makes the speech chain a closed-loop process.

The complexity of the speech communication process outlined above makes it desirable to divide the entire process into modular stages or levels for scientific studies. A common division of the direct information transfer stages of the speech process, which this book is mainly concerned with, is as follows:

- *Linguistic level*: At this highest level of speech communication, the speaker forms the linguistic concept or message to be conveyed to the listener. That is, the speaker decides to say something linguistically meaningful. This process takes place in the language center(s) of speaker's brain. The basic form of the linguistic message is words, which are organized into sentences according to syntactic constraints. Words are in turn composed of syllables constructed from phonemes or segments, which are further composed of phonological features. At this linguistic level, language is represented in a discrete or symbolic form.

- *Physiological level*: Motor program and articulatory muscle movement are involved at this level of speech generation. The speech motor program takes the instructions, specified by the segments and features formed at the linguistic level, on how the speech sounds are to be produced by the articulatory muscle (i.e., articulators) movement over time. Physiologically, the motor program executes itself by issuing time-varying commands imparting continuous motion to the articulators including the lips, tongue, larynx, jaw, and velum, etc. This process involves coordination among various articulators with different limitations in the movement speed, and it also involves constant corrective feedback. The central scientific issue at this level is how the transformation is accomplished from the discrete linguistic representation to the continuous articulators' movement or dynamics. This is sometimes referred to as the problem of interface between phonology and phonetics.

- *Acoustic level*: As a result of the articulators' movements, acoustic air stream emerges from the lungs, and passes through the vocal cords where a phonation type is developed. The time-varying sound sources created in this way are then filtered by the time-varying acoustic cavities shaped by the moving articulators in the vocal tract. The dynamics of this filter can be mathematically represented and approximated by the changing vocal tract area function over time for many practical purposes. The speech information at the acoustic level is in the form of dynamic sound pattern after this filtering process. The sound wave radiated from the lips (and in some cases from the nose and through the tissues of the face) is the most accessible element of the multiple-level speech process for practical applications. For example, this speech sound wave may be easily picked by a microphone and be converted to analog or digital electronic form for storage or transmission. The electronic form of speech sounds makes it possible to transport them thousands of miles away without loss of fidelity. And computerized speech recognizers gain access to speech data also primarily in the electronic form of the original acoustic sound wave.

- *Auditory and perceptual level*: During human speech communication, the speech sound generated at the acoustic level above impinges upon the eardrums of a listener, where it is first converted to mechanical motion via the ossicles of the middle ear, then to fluid pressure waves in the medium bathing the basilar membrane of the inner ear invoking traveling waves. This finally excites hair cells' electrical, mechanical, and biochemical activities, causing firings in some 30,000 human auditory nerve fibers. These various stages of the processing carry out some nonlinear form of frequency analysis, with the analysis results in the form of dynamic spatial–temporal neural response patterns. The dynamic spatial–temporal neural responses are then sent to higher processing centers in the brain, including the brainstem centers, the thalamus, and the primary auditory cortex. The speech representation in the primary auditory cortex (with a high degree of plasticity) appears to be in the form of multiscale and jointly spectro-temporally modulated patterns. For the listener to extract the linguistic content of speech, a process that we call speech perception or *decoding*, it is necessary to identify the segments and features that underlie the sound pattern based on the speech representation in the primary auditory cortex. The decoding process may be aided by some type of analysis-by-synthesis strategies that make use of general knowledge of the dynamic processes at the physiological and acoustic levels of the speech chain as the "encoder" device for the intended linguistic message.

At all the four levels of the speech communication process above, dynamics play a central role in shaping the linguistic information transfer. At the linguistic level, the dynamics are discrete and symbolic, as is the phonological representation. That is, the discrete phonological symbols (segments or features) change their identities at various points of time in a speech utterance, and no quantitative (numeric) degree of change and precise timing are observed. This can be considered as a weak form of dynamics. In contrast, the articulatory dynamics at the physiological level, and the consequent dynamics at the acoustic level, are of a strong form in that the numerically quantifiable temporal characteristics of the articulator movements and of the acoustic parameters are essential for the trade-off between overcoming the physiological limitations for setting the articulators' movement speed and efficient encoding of the phonological symbols. At the auditory level, the importance of timing in the auditory nerve's firing patterns and in the cortical responses in coding speech has been well known. The dynamic patterns in the aggregate auditory neural responses to speech sounds in many ways reflect the dynamic patterns in the speech signal, e.g., time-varying spectral prominences in the speech signal. Further, numerous types of auditory neurons are equipped with special mechanisms (e.g., adaptation and onset-response properties) to enhance the dynamics and information contrast in the acoustic signal. These properties are especially useful for detecting certain special speech events and for identifying temporal "landmarks" as a prerequisite for estimating the phonological features relevant to consonants [2, 3].

Often, we use our intuition to appreciate speech dynamics—as we speak, we sense the motions of speech articulators and the sounds generated from these motions as continuous flow. When we call this continuous flow of speech organs and sounds as speech dynamics, then we use them in a narrow sense, ignoring their linguistic and perceptual aspects.

As is often said, timing is of essence in speech. The dynamic patterns associated with articulation, vocal tract shaping, sound acoustics, and auditory response have the key property that the timing axis in these patterns is adaptively plastic. That is, the timing plasticity is flexible but not arbitrary. Compression of time in certain portions of speech has a significant effect in speech perception, but not so for other portions of the speech. Some compression of time, together with the manipulation of the local or global dynamic pattern, can change perception of the style of speaking but not the phonetic content. Other types of manipulation, on the other hand, may cause very different effects. In speech perception, certain speech events, such as labial stop bursts, flash extremely quickly over as short as 1–3 ms while providing significant cues for the listener to identify the relevant phonological features. In contrast, for other phonological features, even dropping a much longer chunk of the speech sound would not affect their identification. All these point to the very special status of time in speech dynamics. The time in speech seems to be quite different from the linear flow of time as we normally experience it in our living world.

Within the speech recognition community, researchers often refer to speech dynamics as differential or regression parameters derived from the acoustic vector sequence (called delta, delta–delta, or "dynamic" features) [4, 5]. From the perspective of the four-level speech chain outlined above, such parameters can at best be considered as an ultra-weak form of speech dynamics. We call them ultra-weak not only because they are confined to the acoustic domain (which is only one of the several stages in the complete speech chain), but also because temporal differentiation can be regarded hardly as a full characterization in the actual dynamics even within the acoustic domain. As illustrated in [2, 6, 7], the acoustic dynamics of speech exhibited in spectrograms have the intricate, linguistically correlated patterns far beyond what the simplistic differentiation or regression can characterize. Interestingly, there have been numerous publications on how the use of the differential parameters is problematic and inconsistent within the traditional pattern recognition frameworks and how one can empirically remedy the inconsistency (e.g., [8]). The approach that we will describe in this book gives the subject of dynamic speech modeling a much more comprehensive and rigorous treatment from both scientific and technological perspectives.

1.2   WHAT ARE MODELS OF SPEECH DYNAMICS?

As discussed above, the speech chain is a highly dynamic process, relying on the coordination of linguistic, articulatory, acoustic, and perceptual mechanisms that are individually dynamic as well. How do we make sense of this complex process in terms of its functional role of speech communication? How do we quantify the special role of speech timing? How do the dynamics relate to the variability of speech that has often been said to seriously hamper automatic speech recognition? How do we put the dynamic process of speech into a quantitative form to enable detailed analyses? How can we incorporate the knowledge of speech dynamics into computerized speech analysis and recognition algorithms? The answers to all these questions require building and applying computational models for the dynamic speech process.

A computational model is a form of mathematical abstraction of the realistic physical process. It is frequently established with necessary simplification and approximation aimed at mathematical or computational tractability. The tractability is crucial in making the mathematical abstraction amenable to computer or algorithmic implementation for practical engineering applications. Applying this principle, we define *models of speech dynamics* in the context of this book as the mathematical characterization and abstraction of the physical speech dynamics. These characterization and abstraction are capable of capturing the essence of time-varying aspects in the speech chain and are sufficiently simplified to facilitate algorithm development and engineering system implementation for speech processing applications. It is highly desirable that the models be developed in statistical terms, so that advanced algorithms can be developed to automatically and optimally determine any parameters in the models from a representative set of training data. Further, it is important that the probability for each speech utterance be efficiently computed under any hypothesized word-sequence transcript to make the speech decoding algorithm development feasible.

Motivated by the multiple-stage view of the dynamic speech process outlined in the preceding section, detailed computational models, especially those for the multiple generative stages, can be constructed from the distinctive feature-based linguistic units to acoustic and auditory parameters of speech. These stages include the following:

- A discrete feature-organization process that is closely related to speech gesture overlapping and represents partial or full phone deletion and modifications occurring pervasively in casual speech;

- a segmental target process that directs the model-articulators up-and-down and front-and-back movements in a continuous fashion;

- the target-guided dynamics of model-articulators movements that flow smoothly from one phonological unit to the next; and

- the static nonlinear transformation from the model-articulators to the measured speech acoustics and the related auditory speech representations.

The main advantage of modeling such detailed multiple-stage structure in the dynamic human speech process is that a highly compact set of parameters can then be used to capture phonetic context and speaking rate/style variations in a unified framework. Using this framework, many important subjects in speech science (such as acoustic/auditory correlates of distinctive features, articulatory targets/dynamics, acoustic invariance, and phonetic reduction) and those in speech technology (such as modeling pronunciation variation, long-span context-dependence representation, and speaking rate/style modeling for recognizer design) that were previously studied separately by different communities of researchers can now be investigated in a unified fashion.

Many aspects of the above multitiered dynamic speech model class, together with its scientific background, have been discussed in [9]. In particular, the feature organization/overlapping process, as is central to a version of computational phonology, has been presented in some detail under the heading of "computational phonology." Also, some aspects of auditory speech representation, limited mainly to the peripheral auditory system's functionalities, have been elaborated in [9] under the heading of "auditory speech processing." This book will treat these topics only lightly, especially considering that both computational phonology and high-level auditory processing of speech are still active ongoing research areas. Instead, this book will concentrate on the following:

- The target-based dynamic modeling that interfaces between phonology and articulation-based phonetics;
- the switching dynamic system modeling that represents the continuous, target-directed movement in the "hidden" articulators and in the vocal tract resonances being closely related to the articulatory structure; and
- the relationship between the "hidden" articulatory or vocal tract resonance parameters to the measurable acoustic parameters, enabling the hidden speech dynamics to be mapped stochastically to the acoustic dynamics that are directly accessible to any machine processor.

In this book, these three major components of dynamic speech modeling will be treated in a much greater depth than in [9], especially in model implementation and in algorithm development. In addition, this book will include comprehensive reviews of new research work since the publication of [9] in 2003.

1.3   WHY MODELING SPEECH DYNAMICS?

What are the compelling reasons for carrying out dynamic speech modeling? We provide the answer in two related aspects. First, scientific inquiry into the human speech code has been relentlessly pursued for several decades. As an essential carrier of human intelligence and knowledge, speech is the most natural form of human communication. Embedded in the speech code are linguistic (and para-linguistic) messages, which are conveyed through the four levels of the speech chain outlined earlier. Underlying the robust encoding and transmission of the linguistic messages are the speech dynamics at all the four levels (in either a strong form or a weak form). Mathematical modeling of the speech dynamics provides one effective tool in the scientific methods of studying the speech chain—observing phenomena, formulating hypotheses, testing the hypotheses, predicting new phenomena, and forming new theories. Such scientific studies help understand why humans speak as they do and how humans exploit redundancy and variability by way of multitiered dynamic processes to enhance the efficiency and effectiveness of human speech communication.

Second, advancement of human language technology, especially that in automatic recognition of natural-style human speech (e.g., spontaneous and conversational speech), is also expected to benefit from comprehensive computational modeling of speech dynamics. Automatic speech recognition is a key enabling technology in our modern information society. It serves human–computer interaction in the most natural and universal way, and it also aids the enhancement of human–human interaction in numerous ways. However, the limitations of current speech recognition technology are serious and well known (e.g., [10–13]). A commonly acknowledged and frequently discussed weakness of the statistical model (hidden Markov model or HMM) underlying current speech recognition technology is the lack of adequate dynamic modeling schemes to provide correlation structure across the temporal speech observation sequence [9, 13, 14]. Unfortunately, due to a variety of reasons, the majority of current research activities in this area favor only incremental modifications and improvements to the existing HMM-based state-of-the-art. For example, while the dynamic and correlation modeling is known to be an important topic, most of the systems nevertheless employ only the ultra-weak form of speech dynamics, i.e., differential or delta parameters. A strong form of dynamic speech modeling presented in this book appears to be an ultimate solution to the problem.

It has been broadly hypothesized that new computational paradigms beyond the conventional HMM as a generative framework are needed to reach the goal of all-purpose recognition technology for unconstrained natural-style speech, and that statistical methods capitalizing on essential properties of speech structure are beneficial in establishing such paradigms. Over the past decade or so, there has been a popular discriminant-function-based and conditional modeling approach to speech recognition, making use of HMMs (as a discriminant function instead of as a generative model) or otherwise [13, 15–19]. This approach has been grounded on the assumption that we do not have adequate knowledge about the realistic speech process, as exemplified by the following quote from [17]: "The reason of taking a discriminant function based approach to classifier design is due mainly to the fact that we lack complete knowledge of the form of the data distribution and training data are inadequate." The special difficulty of acquiring such distributional speech knowledge lies in the sequential nature of the data with a variable and high dimensionality. This is essentially the problem of dynamics in the speech data. As we gradually fill in such knowledge while pursing research in dynamic speech modeling, we will be able to bridge the gap between the discriminative paradigm and the generative modeling one, but with a much higher performance level than the systems at present. This dynamic speech modeling approach can enable us to "put speech science back into speech recognition" instead of treating speech recognition as a generic, loosely constrained pattern recognition problem. In this way, we are able to develop models "that really model speech," and such models can be expected to provide an opportunity to lay a foundation of the next-generation speech recognition technology.

1.4 OUTLINE OF THE BOOK

After the introduction chapter, the main body of this book consists of four chapters. They cover theory, algorithms, and applications of dynamic speech models and survey in a comprehensive manner the research work in this area spanning over past 20 years or so. In Chapter 2, a general framework for modeling and for computation is presented. It provides the design philosophy for dynamic speech models and outlines five major model components, including phonological construct, articulatory targets, articulatory dynamics, acoustic dynamics, and acoustic distortions. For each of these components, relevant speech science literatures are discussed, and general mathematical descriptions are developed with needed approximations introduced and justified. Dynamic Bayesian networks are exploited to provide a consistent probabilistic language for quantifying the statistical relationships among all the random variables in the dynamic speech models, including both within-component and cross-component relationships.

Chapter 3 is devoted to a comprehensive survey of many different types of statistical models for speech dynamics, from the simple ones that focus on only the observed acoustic patterns to the more advanced ones that represent the dynamics internal to the surface acoustic domain and represent the relationship between these "hidden" dynamics and the observed acoustic dynamics. This survey classifies the existing models into two main categories—acoustic dynamic models and hidden dynamic models, and provides a unified perspective viewing these models as having different degrees of approximation to the realistic multicomponent overall speech chain. Within each of these two main model categories, further classification is made depending on whether the dynamics are mathematically defined with or without temporal recursion. Consequences of this difference in the algorithm development are addressed and discussed.

Chapters 4 and 5 present two types of hidden dynamic models that are best developed to date as reported in the literature, with distinct model classes and distinct approximation and implementation strategies. They exemplify the state-of-the-arts in the research area of dynamic speech modeling. The model described in Chapter 4 uses discretization of the hidden dynamic variables to overcome the original difficulty of intractability in algorithms for parameter estimation and for decoding the phonological states. Modeling accuracy is inherently limited to the discretization precision, and the new computation difficulty arising from the large discretization levels due to multi-dimensionality in the hidden dynamic variables is addressed by a greedy optimization technique. Except for these two approximations, the parameter estimation and decoding algorithms developed and described in this chapter are based on rigorous EM and dynamic programming techniques. Applications of this model and the related algorithms to the problem of automatic hidden vocal tract resonance tracking are presented, where the estimates are for the discretized hidden resonance values determined by the dynamic programming technique for decoding based on the EM-trained model parameters.

The dynamic speech model presented in Chapter 5 maintains the continuous nature in the hidden dynamic values, and uses an explicit temporal function (i.e., defined nonrecursively) to represent the hidden dynamics or "trajectories." The approximation introduced to overcome the original intractability problem is made by iteratively refining the boundaries associated with the discrete phonological units while keeping the boundaries fixed when carrying out parameter estimation. We show computer simulation results that demonstrate the desirable model behavior in characterizing coarticulation and phonetic reduction. Applications to phonetic recognition are also presented and analyzed.

CHAPTER 2

A General Modeling and Computational Framework

The main aim of this chapter is to set up a general modeling and computational framework, based on the modern mathematical tool called dynamic Bayesian networks (DBN) [20, 21], and to establish general forms of the multistage dynamic speech model outlined in the preceding chapter. The overall model presented within this framework is comprehensive in nature, covering all major components in the speech generation chain—from the multitiered, articulation-based phonological construct (top) to the environmentally distorted acoustic observation (bottom). The model is formulated in specially structured DBN, in which the speech dynamics at separate levels of the generative chain are represented by distinct (but dependent) discrete and continuous state variables and by their characteristic temporal evolution.

Before we present the model and the associated computational framework, we first provide a general background and literature review.

2.1 BACKGROUND AND LITERATURE REVIEW

In recent years, the research community in automatic speech recognition has started attacking the difficult problem in the research field—conversational and spontaneous speech recognition (e.g., [12, 16, 22–26]). This new endeavor has been built upon the dramatic progress in speech technology achieved over the past two decades or so [10, 27–31]. While the progress has already created many state-of-the-art speech products, ranging from free-text dictation to wireless access of information via voice, the conditions under which the technology works well are highly limited. The next revolution in speech recognition technology should enable machines to communicate with humans in a natural and unconstrained way. To achieve this challenging goal, some of researchers (e.g., [3, 10, 11, 13, 20, 22, 32–39]) believe that the severe limitations of the HMM should be overcome and novel approaches to representing key aspects of the human speech process are highly desirable or necessary. These aspects, many of which are of a dynamic nature, have been largely missing in the conventional hidden Markov model (HMM) based framework. Towards this objective, one specific strategy would be to place appropriate dynamic structure on the speech model that allows for the kinds of variations observed in conversational speech. Furthermore, enhanced computational methods, including learning and inference techniques, will also be needed based on new or extended models beyond the HMM.

It has been well known that many assumptions in the current HMM framework are inconsistent with the underlying mechanisms, as well as with the surface acoustic observations, of the dynamic speech process (e.g., [13, 14]). However, a number of approaches, notably known as the stochastic segment model, segmental HMM, and trended trajectory models [14, 40, 41], which are intended to specifically overcome some of these inconsistent assumptions, have not delivered significant and consistent performance improvements, especially in large speech recognition systems. Most of these approaches aimed mainly at overcoming the HMM's assumption of conditional IID (independent and identical distribution conditioned on the HMM state sequence). Yet the inconsistency between the HMM assumptions and the properties of the realistic dynamic speech process goes far beyond this "local" IID limitation. It appears necessary not just to empirically fix one weak aspect of the HMM or another, but rather to develop the new computational machinery that directly incorporates key dynamic properties of the human speech process. One natural (but challenging) way of incorporating such knowledge is to build comprehensive statistical generative models for the speech dynamics. Via the use of Bayes theorem, the statistical generative models explicitly give the posterior probabilities for different speech classes, enabling effective decision rules to be constructed in a speech recognizer. In discriminative modeling approaches, on the other hand, where implicit computation of the posterior probabilities for speech classes is carried out, it is generally much more difficult to systematically incorporate knowledge of the speech dynamics.

Along the direction of generative modeling, many researchers have, over recent years, been proposing and pursuing research that extensively explores the dynamic properties of speech in various forms and at various levels of the human speech communication chain (e.g., [24, 32–34, 42–49]). Some approaches have advocated the use of the multitiered feature-based phonological units, which control human speech production and are typical of human lexical representation (e.g., [11, 50–52]). Other approaches have emphasized the functional significance of abstract, "task" dynamics in speech production and recognition (e.g., [53, 54]). Yet other approaches have focused on the dynamic aspects in the speech process, where the dynamic object being modeled is in the space of surface speech acoustics, rather than in the space of the intermediate, production-affiliated variables that are internal to the direct acoustic observation (e.g., [14, 26, 55–57]).

Although dynamic modeling has been a central focus of much recent work in speech recognition, the dynamic object being modeled, either in the space of "task" variables or of acoustic variables, does not, and potentially may not be able to, directly take into account the many important properties in realistic articulatory dynamics. Some earlier proposals and empirical methods for modeling pseudo-articulatory dynamics or abstract hidden dynamics for the purpose of speech recognition can be found in [42, 44–46, 58, 59]. In these studies, the dynamics of a set of pseudo-articulators are realized either by filtering from sequentially arranged, phoneme-specific target positions or by applying trajectory-smoothness constraints. Due to the simplistic nature in the use of the pseudo-articulators, one important property of human speech production, compensatory articulation, could not be taken into account, because it would require modeling correlations among target positions of a set of articulators. This has reduced the power of such models for potentially successful use in speech recognition.

To incorporate essential properties in human articulatory dynamics—including compensatory articulation, target-directed behavior, and flexibly constrained dynamics due to biomechanical properties of different articulatory organs—in a statistical generative model of speech, it appears necessary to use essential properties of realistic multidimensional articulators. Previous attempts using the pseudo-articulators did not incorporate most of such essential properties. Because much of the acoustic variation observed in speech that makes speech recognition difficult can be attributed to articulatory phenomena, and because articulation is one key component in the closed-loop human speech communication chain, it is highly desirable to develop an explicit articulation-motivated dynamic model and to incorporate it into a comprehensive generative model of the dynamic speech process.

The comprehensive generative model of speech and the associated computational framework discussed in this chapter consists of a number of key components that are centered on articulatory dynamics. A general overview of this multicomponent model is provided next, followed by details of the individual model components including their mathematical descriptions and their DBN representations.

2.2   MODEL DESIGN PHILOSOPHY AND OVERVIEW

Spontaneous speech (e.g., natural voice mails and lectures) and speech of verbal conversations among two or more speakers (e.g., over the telephone or in meetings) are pervasive forms of human communication. If a computer system can be constructed to automatically decode the linguistic messages contained in spontaneous and conversational speech, one will have vast opportunities for the applications of speech technology.

What characterizes spontaneous and conversational speech is its *casual style* nature. The casual style of speaking produces two key consequences that make the acoustics of spontaneous and conversational speech significantly differ from that of the "read-style" speech: phonological reorganization and phonetic reduction. First, in casual speech, which is called "hypo-articulated" speech in [60], phonological reorganization occurs where the relative timing or phasing relationship across different articulatory feature/gesture dimensions in the "orches-trated" feature strands are modified. One obvious manifestation of this modification is that the more casual or relaxed the speech style is, the greater the overlapping across the feature/gesture dimensions becomes. Second, phonetic reduction occurs where articulatory targets as phonetic correlates to the phonological units may shift towards a more neutral position due to the use of reduced articulatory efforts. Phonetic reduction also manifests itself by pulling the realized articulatory trajectories further away from reaching their respective targets due to physical inertia constraints in the articulatory movements. This occurs within generally shorter time duration in casual-style speech than in the read-style speech.

It seems difficult for the HMM systems to provide effective mechanisms to embrace the huge, new acoustic variability in casual, spontaneous, and conversational speech arising either from phonological organization or from phonetic reduction. Importantly, the additional variability due to phonetic reduction is scaled continuously, resulting in phonetic confusions in a predictable manner. (See Chapter 5 for some detailed computation simulation results pertaining to such prediction.) Due to this continuous variability scaling, very large amounts of (labeled) speech data would be needed. Even so, they can only partly capture the variability when no structured knowledge about phonetic reduction and about its effects on speech dynamic patterns is incorporated into the speech model underlying spontaneous and conversational speech-recognition systems.

The general design philosophy of the mathematical model for the speech dynamics described in this chapter is based on the desire to integrate the structured knowledge of both phonological reorganization and phonetic reduction. To fully describe this model, we break up the model into several interrelated components, where the output, expressed as the probability distribution, of one component serves as the input to the next component in a "generative" spirit. That is, we characterize each model component as a joint probability distribution of both input and output sequences, where both the sequences may be hidden. The top-level component is the phonological model that specifies the discrete (symbolic) pronunciation units of the intended linguistic message in terms of multitiered, overlapping articulatory features. The first intermediate component consists of articulatory control and target, which provides the interface between the discrete phonological units to the continuous phonetic variable and which represents the "ideal" articulation and its inherent variability if there were no physical constraints in articulation. The second intermediate component consists of articulatory dynamics, which explicitly represents the physical constraints in articulation and gives the output of "actual" trajectories in the articulatory variables. The bottom component would be the process of speech acoustics being generated from the vocal tract whose shape and excitation are determined by the articulatory variables as the output of the articulatory dynamic model. However, considering that such a clean signal is often subject to one form of acoustic distortion or another before being processed by a speech recognizer, and further that the articulatory behavior and the subsequent speech dynamics in acoustics may be subject to change when the acoustic distortion becomes severe, we complete the comprehensive model by adding the final component of acoustic distortion with feedback to the higher level component describing articulatory dynamics.

2.3   MODEL COMPONENTS AND THE COMPUTATIONAL FRAMEWORK

In a concrete form, the generative model for speech dynamics, whose design philosophy and motivations have been outlined in the preceding section, consists of the hierarchically structured components of

1. multitiered phonological construct (nonobservable or hidden; discrete valued);

2. articulatory targets (hidden; continuous-valued);

3. articulatory dynamics (hidden; continuous);

4. acoustic pattern formation from articulatory variables (hidden; continuous); and

5. distorted acoustic pattern formation (observed; continuous).

In this section, we will describe each of these components and their design in some detail. In particular, as a general computational framework, we provide the DBN representation for each of the above model components and for their combination.

2.3.1   Overlapping Model for Multitiered Phonological Construct

Phonology is concerned with sound patterns of speech and with the nature of the discrete or symbolic units that shapes such patterns. Traditional theories of phonology differ in the choice and interpretation of the phonological units. Early distinctive feature-based theory [61] and subsequent autosegmental, feature-geometry theory [62] assumed a rather direct link between phonological features and their phonetic correlates in the articulatory or acoustic domain. Phonological rules for modifying features represented changes not only in the linguistic structure of the speech utterance, but also in the phonetic realization of this structure. This weakness has been recognized by more recent theories, e.g., articulatory phonology [63], which emphasize the importance of accounting for phonetic levels of variation as distinct from those at the phonological levels.

In the phonological model component described here, it is assumed that the linguistic function of phonological units is to maintain linguistic contrasts and is separate from phonetic implementation. It is further assumed that the phonological unit sequence can be described mathematically by a discrete-time, discrete-state, multidimensional homogeneous Markov chain. How to construct sequences of symbolic phonological units for any arbitrary speech utterance and how to build them into an appropriate Markov state (i.e., phonological state) structure are two key issues in the model specification. Some earlier work on effective methods of constructing such overlapping units, either by rules or by automatic learning, can be found in [50, 59, 64–66]. In limited experiments, these methods have proved effective for coarticulation modeling in the HMM-like speech recognition framework (e.g., [50, 65]).

Motivated by articulatory phonology [63], the asynchronous, feature-based phonological model discussed here uses multitiered articulatory features/gestures that are temporally overlapping with each other in separate tiers, with learnable relative-phasing relationships. This contrasts with most existing speech-recognition systems where the representation is based on phone-sized units with one single tier for the phonological sequence acting as "beads-on-a-string." This contrast has been discussed in some detail in [11] with useful insight.

Mathematically, the L-tiered, overlapping model can be described by the "factorial" Markov chain [51, 67], where the state of the chain is represented by a collection of discrete-component state variables for each time frame t:

$$\mathbf{s}_t = s_t^{(1)}, \ldots, s_t^{(l)}, \ldots, s_t^{(L)}.$$

Each of the component states can take $K^{(l)}$ values. In implementing this model for American English, we have $L = 5$, and the five tiers are Lips, Tongue Blade, Tongue Body, Velum, and Larynx, respectively. For "Lips" tier, we have $K^{(1)} = 6$ for six possible linguistically distinct Lips configurations, i.e., those for /b/, /r/, /sh/, /u/, /w/, and /o/. Note that at this phonological level, the difference among these Lips configurations is purely symbolic. The numerical difference is manifested in different articulatory target values at lower phonetic level, resulting ultimately in different acoustic realizations. For the remaining tiers, we have $K^{(2)} = 6$, $K^{(3)} = 17$, $K^{(4)} = 2$, and $K^{(5)} = 2$.

The state–space of this factorial Markov chain consists of all $K_L = K^{(1)} \times K^{(2)} \times K^{(3)} \times K^{(4)} \times K^{(5)}$ possible combinations of the $s_t^{(l)}$ state variables. If no constraints are imposed on the state transition structure, this would be equivalent to the conventional one-tiered Markov chain with a total of K_L states and a $K_L \times K_L$ state transition matrix. This would be an uninteresting case since the model complexity is exponentially (or factorially) growing in L. It would also be unlikely to find any useful phonological structure in this huge Markov chain. Further, since all the phonetic parameters in the lower level components of the overall model (to be discussed shortly) are conditioned on the phonological state, the total number of model parameters would be unreasonably large, presenting a well-known sparseness difficulty for parameter learning.

Fortunately, rich sources of phonological and phonetic knowledge are available to constrain the state transitions of the above factorial Markov chain. One particularly useful set of constraints come directly from the phonological theories that motivated the construction of this model. Both autosegmental phonology [62] and articulatory phonology [63] treat the different tiers in the phonological features as being largely independent of each other in their

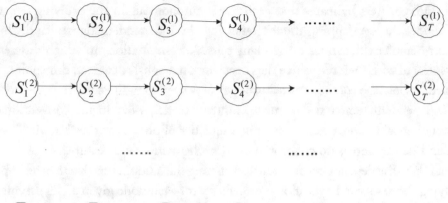

FIGURE 2.1: A dynamic Bayesian network (DBN) for a constrained factorial Markov chain as a probabilistic model for an L-tiered overlapping phonological model based on articulatory features/gestures. The constrained transition structure in the factorial Markov chain makes different tiers in the phonological features independent of each other in their evolving dynamics. This gives rise to parallel streams, $s^{(l)}, l = 1, 2, \ldots, L$, of the phonological features in their associated articulatory dimensions

evolving dynamics. This thus allows the *a priori* decoupling among the L tiers:

$$P(\mathbf{s}_t|\mathbf{s}_{t-1}) = \prod_{l=1}^{L} P(s_t^{(l)}|s_{t-1}^{(l)}).$$

The transition structure of this constrained (uncoupling) factorial Markov chain can be parameterized by L distinct $K^{(l)} \times K^{(l)}$ matrices. This is significantly simpler than the original $K_L \times K_L$ matrix as in the unconstrained case.

Fig. 2.1 shows a dynamic Bayesian network (DBN) for a factorial Markov chain with the constrained transition structure. A Bayesian network is a graphical model that describes dependencies and conditional independencies in the probabilistic distributions defined over a set of random variables. The most interesting class of Bayesian networks, as relevant to speech modeling, is the DBN specifically aimed at modeling time series data or symbols such as speech acoustics, phonological units, or a combination of them. For the speech data or symbols, there are causal dependencies between random variables in time and they are naturally suited for the DBN representation.

In the DBN representation of Fig. 2.1 for the L-tiered phonological model, each node represents a component phonological feature in each tier as a discrete random variable at a particular discrete time. The fact that there is no dependency (lacking arrows) between the nodes in different tiers indicates that each tier is autonomous in the evolving dynamics. We call this model the overlapping model, reflecting the independent dynamics of the features at different tiers. The dynamics cause many possible combinations in which different feature values associated with their respective tiers occur simultaneously at a fixed time point. These are determined by how the component features/gestures overlap with each other as a consequence of their independent temporal dynamics. Contrary to this view, in the conventional phone-based phonological model, there is only one single tier of phones as the "bundled" component features, and hence there is no concept of overlapping component features.

In a DBN, the dependency relationships among the random variables can be implemented by specifying the associated conditional probability for each node given all its parents. Because of the decoupled structure across the tiers as shown in Fig. 2.1, the horizontal (temporal) dependency is the only dependency that exists for the component phonological (discrete) states. This dependency can be specified by the Markov chain transition probability for each separate tier, l, defined by

$$P\big(s_t^{(l)} = j | s_{t-1}^{(l)} = i\big) = a_{ij}^{(l)}. \tag{2.1}$$

2.3.2   Segmental Target Model

After a phonological model is constructed, the process for converting abstract phonological units into their phonetic realization needs to be specified. The key issue here is whether the invariance in the speech process is more naturally expressed in the articulatory or the acoustic/auditory domain. A number of theories assumed a direct link between abstract phonological units and physical measurements. For example, the "quantal theory" [68] proposed that phonological features possessed invariant acoustic (or auditory) correlates that could be measured directly from the speech signal. The "motor theory" [69] instead proposed that articulatory properties are directly associated with phonological symbols. No conclusive evidence supporting either hypothesis has been found without controversy [70].

In the generative model of speech dynamics discussed here, one commonly held view in phonetics literature is adopted. That is, discrete phonological units are associated with a temporal segmental sequence of phonetic targets or goals [71–75]. In this view, the function of the articulatory motor control system is to achieve such targets or goals by manipulating the articulatory organs according to some control principles subject to the articulatory inertia and possibly minimal-energy constraints [60].

Compensatory articulation has been widely documented in the phonetics literature where trade-offs between different articulators and nonuniqueness in the articulatory–acoustic mapping allow for the possibility that many different articulatory target configurations may be able to "equivalently" realize the same underlying goal. Speakers typically choose a range of possible targets depending on external environments and their interactions with listeners [60, 70, 72, 76, 77]. To account for compensatory articulation, a complex phonetic control strategy need to be adopted. The key modeling assumptions adopted here regarding such a strategy are as follows. First, each phonological unit is correlated to a number of phonetic parameters. These measurable parameters may be acoustic, articulatory, or auditory in nature, and they can be computed from some physical models for the articulatory and auditory systems. Second, the region determined by the phonetic correlates for each phonological unit can be mapped onto an articulatory parameter space. Hence, the target distribution in the articulatory space can be determined simply by stating what the phonetic correlates (formants, articulatory positions, auditory responses, etc.) are for each of the phonological units (many examples are provided in [2]), and by running simulations in detailed articulatory and auditory models. This particular proposal for using the joint articulatory, acoustic, and auditory properties to specify the articulatory control in the domain of articulatory parameters was originally proposed in [59, 78]. Compared with the traditional modeling strategy for controlling articulatory dynamics [79] where the sole articulatory goal is involved, this new strategy appears more appealing not only because of the incorporation of the perceptual and acoustic elements in the specification of the speech production goal, but also because of its natural introduction of statistical distributions at the relatively high level of speech production.

A convenient mathematical representation for the distribution of the articulatory target vector $\mathbf{t}$ follows a multivariate Gaussian distribution, denoted by

$$\mathbf{t} \sim \mathcal{N}(\mathbf{t}; \mathbf{m}(\mathbf{s}), \Sigma(\mathbf{s})), \tag{2.2}$$

where $\mathbf{m}(\mathbf{s})$ is the mean vector associated with the composite phonological state $\mathbf{s}$, and the covariance matrix $\Sigma(\mathbf{s})$ is nondiagonal. This allows for the correlation among the articulatory vector components. Because such a correlation is represented for the articulatory target (as a random vector), compensatory articulation is naturally incorporated in the model.

Since the target distribution, as specified in Eq. (2.2), is conditioned on a specific phonological unit (e.g., a bundle of overlapped features represented by the composite state $\mathbf{s}$ consisting of component feature values in the factorial Markov chain of Fig. 2.1), and since the target does not switch until the phonological unit changes, the statistics for the temporal sequence of the target process follows that of a segmental HMM [40].

For the single-tiered ($L = 1$) phonological model (e.g., phone-based model), the segmental HMM for the target process will be the same as that described in [40], except the output is no longer the acoustic parameters. The dependency structure in this segmental HMM as the combined one-tiered phonological model and articulatory target model can be illustrated in the DBN of Fig. 2.2. We now elaborate on the dependencies in Fig. 2.2. The

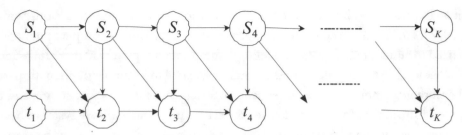

FIGURE 2.2: DBN for a segmental HMM as a probabilistic model for the combined one-tiered phonological model and articulatory target model. The output of the segmental HMM is the target vector, **t**, constrained to be constant until the discrete phonological state, s, changes its value

output of this segmental HMM is the random articulatory target vector $\mathbf{t}(k)$ that is constrained to be constant until the phonological state switches its value. This segmental constraint for the dynamics of the random target vector $\mathbf{t}(k)$ represents the adopted articulatory control strategy that the goal of the motor system is to try to maintain the articulatory target's position (for a fixed corresponding phonological state) by exerting appropriate muscle forces. That is, although random, $\mathbf{t}(k)$ remains fixed until the phonological state s_k switches. The switching of target $\mathbf{t}(k)$ is synchronous with that of the phonological state, and only at the time of switching, is $\mathbf{t}(k)$ allowed to take a new value according to its probability density function. This segmental constraint can be described mathematically by the following conditional probability density function:

$$p[\mathbf{t}(k)|s_k, s_{k-1}, \mathbf{t}(k-1)] = \begin{cases} \delta[\mathbf{t}(k) - \mathbf{t}(k-1)] & \text{if } s_k = s_{k-1}, \\ \mathcal{N}(\mathbf{t}(k); \mathbf{m}(s_k), \Sigma(s_k)) & \text{otherwise.} \end{cases}$$

This adds the new dependencies of random vector of $\mathbf{t}(k)$ on s_{k-1} and $\mathbf{t}(k-1)$, in addition to the obvious direct dependency on s_k, as shown in Fig. 2.2.

Generalizing from the one-tiered phonological model to the multitiered one as discussed earlier, the dependency structure in the "segmental factorial HMM" as the combined multitiered phonological model and articulatory target model has the DBN representation in Fig. 2.3. The key conditional probability density function (PDF) is similar to the above segmental HMM except that the conditioning phonological states are the composite states ($\mathbf{s}_k$ and $\mathbf{s}_{k-1}$) consisting of a collection of discrete component state variables:

$$p[\mathbf{t}(k)|\mathbf{s}_k, \mathbf{s}_{k-1}, \mathbf{t}(k-1)] = \begin{cases} \delta[\mathbf{t}(k) - \mathbf{t}(k-1)] & \text{if } \mathbf{s}_k = \mathbf{s}_{k-1}, \\ \mathcal{N}(\mathbf{t}(k); \mathbf{m}(\mathbf{s}_k), \Sigma(\mathbf{s}_k)) & \text{otherwise.} \end{cases}$$

Note that in Figs. 2.2 and 2.3 the target vector $\mathbf{t}(k)$ is defined in the same space as that of the physical articulator vector (including jaw positions, which do not have direct phonological

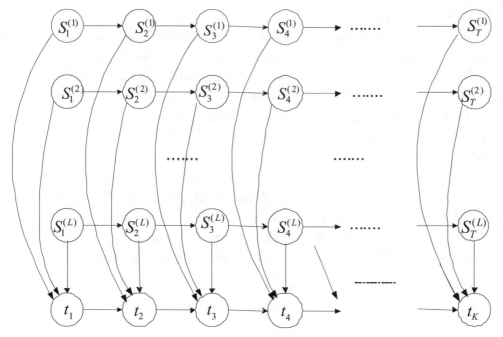

FIGURE 2.3: DBN for a segmental factorial HMM as a combined multitiered phonological model and articulatory target model

connections). And compensatory articulation can be represented directly by the articulatory target distributions with a nondiagonal covariance matrix for component correlation. This correlation shows how the various articulators can be jointly manipulated in a coordinated manner to produce the same phonetically implemented phonological unit.

An alternative model for the segmental target model, as proposed in [33] and called the "target dynamic" model, uses vocal tract constrictions (degrees and locations) instead of articulatory parameters as the target vector, and uses a geometrically defined nonlinear relationship (e.g., [80]) to map one vector of vocal tract constrictions into a region (with a probability distribution) of the physical articulatory variables. In this case, compensatory articulation can also be represented by the distributional region of the articulatory vectors induced indirectly by any fixed vocal tract constriction vector.

The segmental factorial HMM presented here is a generalization of the segmental HMM proposed originally in [40]. It is also a generalization of the factorial HMM that has been developed from the machine learning community [67] and been applied to speech recognition [51]. These generalizations are necessary because the output of our component model (not the full model) is physically the time-varying articulatory targets as random sequences, rather than the random acoustic sequences as in the segmental HMM or the factorial HMM.

2.3.3   Articulatory Dynamic Model

Due to the difficulty in knowing how the conversion of higher-level motor control into articulator movement takes place, a simplifying assumption is made for the new model component discussed in this subsection. That is, we assume that at the functional level the combined (nonlinear) control system and articulatory mechanism behave as a linear dynamic system. This combined system attempts to track the control input, equivalently represented by the articulatory target, in the physical articulatory parameter space. Articulatory dynamics can then be approximated as the response of a linear filter driven by a random target sequence as represented by a segmental factorial HMM just described. The statistics of the random target sequence approximate those of the muscle forces that physically drives motions of the articulators. (The output of this hidden articulatory dynamic model then produces a time-varying vocal tract shape that modulates the acoustic properties of the speech signal.)

The above simplifying assumption then reduces the generally intractable nonlinear state equation,

$$\mathbf{z}(k+1) = \mathbf{g}_s[\mathbf{z}(k), \mathbf{t}_s, \mathbf{w}(k)],$$

into the following mathematically tractable, linear, first-order autoregressive (AR) model:

$$\mathbf{z}(k+1) = \mathbf{A}_s\mathbf{z}(k) + \mathbf{B}_s\mathbf{t}_s + \mathbf{w}(k), \qquad (2.3)$$

where $\mathbf{z}$ is the n-dimensional real-valued articulatory-parameter vector, $\mathbf{w}$ is the IID and Gaussian noise, $\mathbf{t}_s$ is the HMM-state-dependent target vector expressed in the same articulatory domain as $\mathbf{z}(k)$, $\mathbf{A}_s$ is the HMM-state-dependent system matrix, and $\mathbf{B}_s$ is a matrix that modifies the target vector. The dependence of $\mathbf{t}_s$ and Φ_s parameters of the above dynamic system on the phonological state is justified by the fact that the functional behavior of an articulator depends both on the particular goal it is trying to implement, and on the other articulators with which it is cooperating in order to produce compensatory articulation.

In order for the modeled articulatory dynamics above to exhibit realistic behaviors, e.g., movement along the target-directed path within each segment and not oscillating within the segment, matrices $\mathbf{A}_s$ and $\mathbf{B}_s$ can be constrained appropriately. One form of the constraint gives rise to the following articulatory dynamic model:

$$\mathbf{z}(k+1) = \Phi_s\mathbf{z}(k) + (\mathbf{I} - \Phi_s)\mathbf{t}_s + \mathbf{w}(k), \qquad (2.4)$$

where $\mathbf{I}$ is the identity matrix. Other forms of the constraint will be discussed in Chapters 4 and 5 of the book for two specific implementations of the general model.

It is easy to see that the constrained linear AR model of Eq. (2.4) has the desirable target-directed property. That is, the articulatory vector $\mathbf{z}(k)$ asymptotically approaches the mean of the target random vector $\mathbf{t}$ for artificially lengthened speech utterances. For natural speech, and especially for conversational speech with a casual style, the generally short duration associated with each phonological state forces the articulatory dynamics to move away from the target of the current state (and towards the target of the following phonological state) long before it reaches the current target. This gives rise to phonetic reduction, and is one key source of speech variability that is difficult to be directly captured by a conventional HMM.

Including the linear dynamic system model of Eq. (2.4), the combined phonological, target, and articulatory dynamic model now has the DBN representation of Fig. 2.4. The new dependency for the continuous articulatory state is specified, on the basis of Eq. (2.4), by the following conditional PDF:

$$p_{\mathbf{z}}[\mathbf{z}(k+1)|\mathbf{z}(k), \mathbf{t}(k), \mathbf{s}_k] = p_{\mathbf{w}}[\mathbf{z}(k+1) - \Phi_{\mathbf{s}_k}\mathbf{z}(k) - (\mathbf{I} - \Phi_{\mathbf{s}_k})\mathbf{t}(k)]. \qquad (2.5)$$

This combined model is a switching, target-directed AR model driven by a segmental factorial HMM.

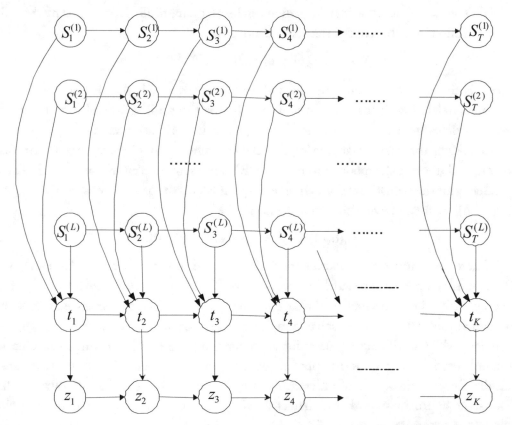

FIGURE 2.4: DBN for a switching, target-directed AR model driven by a segmental factorial HMM. This is a combined model for multitiered phonology, target process, and articulatory dynamics

2.3.4   Functional Nonlinear Model for Articulatory-to-Acoustic Mapping

The next component in the overall model of speech dynamics moves the speech generative process from articulation down to distortion-free speech acoustics. While a truly consistent framework to accomplish this based on explicit knowledge of speech production should include detailed mechanisms for articulatory-to-acoustic generation, this becomes impractical due to difficulties in modeling learning and excessive computational requirements. Again, we make the simplifying assumptions that the articulatory and acoustic states of the vocal tract can be adequately described by low-order vectors of variables, where the articulatory state variables represent respectively the relative positions of the major articulators, and the acoustic state variables represent corresponding time-averaged spectral-like parameters computed from the acoustic signal. If an appropriate time scale is chosen, the relationship between the articulatory and acoustic representations can be modeled by a static memoryless transformation, converting a vector of articulatory parameters into a vector of acoustic ones. This assumption appears reasonable for a vocal tract with about 10-ms reverberation time.

This static memoryless transformation can be mathematically represented by the following "observation" equation in the state–space model:

$$\mathbf{o}(k) = \mathbf{h}[\mathbf{z}(k)] + \mathbf{v}(k), \tag{2.6}$$

where $\mathbf{o}$ is the m-dimensional real-valued observation vector, $\mathbf{v}$ is the IID observation noise vector uncorrelated with the state noise $\mathbf{w}$, and $\mathbf{h}[\cdot]$ is the static memoryless transformation from the articulatory vector to its corresponding acoustic observation vector.

Including this static mapping model, the combined phonological, target, articulatory dynamic, and the acoustic model now has the DBN representation shown in Fig. 2.5. The new dependency for the acoustic random variables is specified, on the basis of "observation" equation in Eq. (2.6), by the following conditional PDF:

$$p_{\mathbf{o}}[\mathbf{o}(k) \,|\, \mathbf{z}(k)] = p_{\mathbf{v}}[\mathbf{o}(k) - \mathbf{h}(\mathbf{z}(k))]. \tag{2.7}$$

There are many ways of choosing the static nonlinear function for $\mathbf{h}[\mathbf{z}]$ in Eq. (2.6), such as using a multilayer perceptron (MLP) neural network. Typically, the analytical forms of nonlinear functions make the associated nonlinear dynamic systems difficult to analyze and make the estimation problems difficult to solve. Simplification is frequently used to gain computational advantages while sacrificing accuracy for approximating the nonlinear functions. One most commonly used technique for the approximation is truncated (vector) Taylor series expansion. If all the Taylor series terms of order two and higher are truncated, then we have the linear Taylor series approximation that is characterized by the Jacobian matrix $\mathbf{J}$ and by the point of Taylor series expansion $\mathbf{z}_0$:

$$\mathbf{h}(\mathbf{z}) \approx \mathbf{h}(\mathbf{z}_0) + \mathbf{J}(\mathbf{z}_0)(\mathbf{z} - \mathbf{z}_0). \tag{2.8}$$

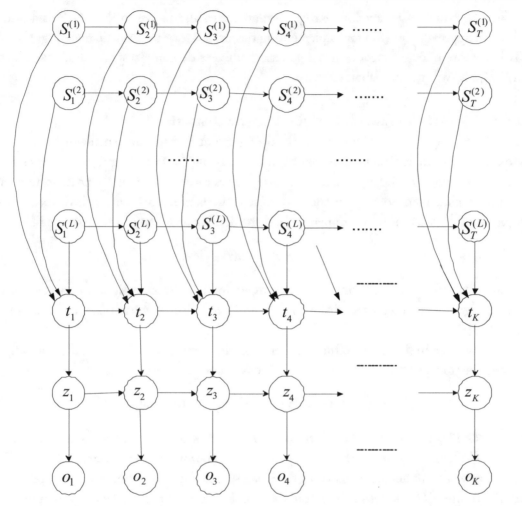

FIGURE 2.5: DBN for a target-directed, switching dynamic system (state–space) model driven by a segmental factorial HMM. This is a combined model for multitiered phonology, target process, articulatory dynamics, and articulatory-to-acoustic mapping

Each element of the Jacobian matrix $\mathbf{J}$ is partial derivative of each vector component of the nonlinear output with respect to each of the input vector components. That is,

$$\mathbf{J}(\mathbf{z}_0) = \frac{\partial \mathbf{h}}{\partial \mathbf{z}_0} = \begin{bmatrix} \frac{\partial h_1(\mathbf{z}_0)}{\partial z_1} & \frac{\partial h_1(\mathbf{z}_0)}{\partial z_2} & \cdots & \frac{\partial h_1(\mathbf{z}_0)}{\partial z_n} \\ \frac{\partial h_2(\mathbf{z}_0)}{\partial z_1} & \frac{\partial h_2(\mathbf{z}_0)}{\partial z_2} & \cdots & \frac{\partial h_2(\mathbf{z}_0)}{\partial z_n} \\ \vdots & & \vdots & \\ \frac{\partial h_m(\mathbf{z}_0)}{\partial z_1} & \frac{\partial h_m(\mathbf{z}_0)}{\partial z_2} & \cdots & \frac{\partial h_m(\mathbf{z}_0)}{\partial z_n} \end{bmatrix}. \tag{2.9}$$

The radial basis function (RBF) is an attractive alternative to the MLP as a universal function approximator [81] for implementing the articulatory-to-acoustic mapping. Use of the RBF for a nonlinear function in the general nonlinear dynamic system model can be found in [82], and will not be elaborated here.

2.3.5   Weakly Nonlinear Model for Acoustic Distortion

In practice, the acoustic vector of speech, $\mathbf{o}(k)$, generated from the articulatory vector $\mathbf{z}(k)$ is subject to acoustic environmental distortion before being "observed" and being processed by a speech recognizer for linguistic decoding. In many cases of interest, acoustic distortion can be accurately characterized by joint additive noise and channel (convolutional) distortion in the signal-sample domain. In this domain, the distortion is linear and can be described by

$$y(t) = o(t) * \hbar(t) + n(t), \tag{2.10}$$

where $y(t)$ is the distorted speech signal sample, modeled as convolution between the clean speech signal sample $o(t)$ and distortion channel's impulse response $\hbar(t)$ plus additive noise sample $n(t)$.

However, in the log-spectrum domain or in the cepstrum domain that is commonly used as the input for speech recognizers, Eq. (2.10) has its equivalent form of (see a derivation in [83])

$$\mathbf{y}(k) = \mathbf{o}(k) + \mathbf{h} + \mathbf{C} \log \left[\mathbf{I} + \exp[\mathbf{C}^{-1}(\mathbf{n}(k) - \mathbf{o}(k) - \mathbf{h})] \right]. \tag{2.11}$$

In Eq. (2.11), $\mathbf{y}(k)$ is the cepstral vector at frame k, when $\mathbf{C}$ is taken as cosine transform matrix. (When $\mathbf{C}$ is taken as the identity matrix, $\mathbf{y}(k)$ becomes a log-spectral vector.) $\mathbf{y}(k)$ now becomes weakly nonlinearly related to the clean speech cepstral vector $\mathbf{o}(k)$, cepstral vector of additive noise $\mathbf{n}(k)$, and cepstral vector of the impulse response of the distortion channel $\mathbf{h}$. Note that according to Eq. (2.11), the relationship between clean and noisy speech cepstral vectors becomes linear (or affine) when the signal-to-noise ratio is either very large or very small.

After incorporating the above acoustic distortion model, and assuming that the statistics of the additive noise changes slowly over time as governed by a discrete-state Markov chain, Fig. 2.6 shows the DBN for the comprehensive generative model of speech from the phonological model to distorted speech acoustics. Intermediate models include the target model, articulatory dynamic model, and clean-speech acoustic model. For clarity, only a one-tiered, rather than multitiered, phonological model is illustrated. [The dependency of the parameters (Φ) of the articulatory dynamic model on the phonological state is also explicitly added.] Note that in Fig. 2.6, the temporal dependency in the discrete noise states N_k gives rise to nonstationarity in the additive noise random vectors $\mathbf{n}_k$. The cepstral vector $\mathbf{h}$ for the distortion channel is assumed not to change over the time span of the observed distorted speech utterance $\mathbf{y}_1, \mathbf{y}_2, \ldots, \mathbf{y}_K$.

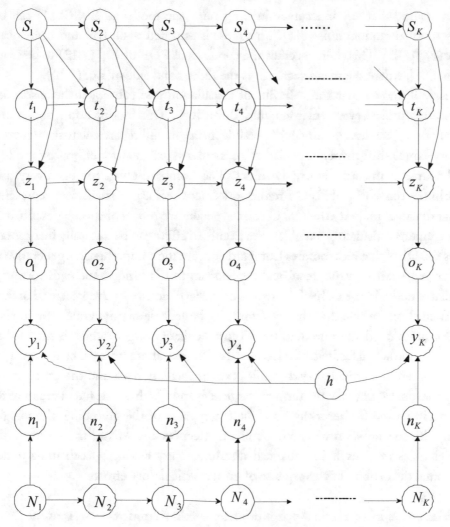

FIGURE 2.6: DBN for a comprehensive generative model of speech from the phonological model to distorted speech acoustics. Intermediate models include the target model, articulatory dynamic model, and acoustic model. For clarity, only a one-tiered, rather than multitiered, phonological model is illustrated. Explicitly added is the dependency of the parameters ($\Phi(s)$) of the articulatory dynamic model on the phonological state

In Fig. 2.6, the dependency relationship for the new variables of distorted acoustic observation $\mathbf{y}(k)$ is specified, on the basis of observation equation 2.11 where $\mathbf{o}(k)$ is specified in Eq. (5), by the following conditional PDF:

$$p(\mathbf{y}(k) \mid \mathbf{o}(k), \mathbf{n}(k), \mathbf{h}) = p_\epsilon(\mathbf{y}(k) - \mathbf{o}(k) - \mathbf{h} - \mathbf{C}\log[\mathbf{I} + \exp(\mathbf{C}^{-1}(\mathbf{n}(k) - \mathbf{o}(k) - \mathbf{h}))]),$$

$$(2.12)$$

where we assume that any inaccuracy in the parametric model of Eq. (2.11) can be represented by residual random noise $\epsilon[k]$. This noise is assumed to be IID and zero-mean Gaussian: $\mathcal{N}[\epsilon(k); \mathbf{0}, \Sigma_\epsilon]$. This then specifies the conditional PDF of Eq. (2.12) to be Gaussian of $\mathcal{N}[\mathbf{y}(k); \mathbf{m}, \Sigma_\epsilon]$, where the mean vector $\mathbf{m}$ is the right-hand side of Eq. (2.11).

It is well known that the behavior of articulation and subsequent acoustics is subject to modification under severe environmental distortions. This modification, sometimes called "Lombard effect," can take a number of possible forms, including articulatory target overshoot, articulatory target shift, hyper-articulation or increased articulatory efforts by modifying the temporal course of the articulatory dynamics. The Lombard effect has been very difficult to represent in the conventional HMM framework since there is no articulatory representation or any similar dynamic property therein. Given the generative model of speech described here that explicitly contains articulatory variables, the Lombard effect can be naturally incorporated. Fig. 2.7 shows the DBN that incorporates Lombard effect in the comprehensive generative model of speech. It is represented by the "feedback" dependency from the noise and h-distortion nodes to the articulator nodes in the DBN. The nature of the feedback may be represented in the form of "hyper-articulation," where the "time constant" in the articulatory dynamic equation is reduced to allow for more rapid attainment of the given articulatory target (which is sampled from the target distribution). The feedback for Lombard effect may alternatively take the form of "target overshoot," where the articulatory dynamics exhibit oscillation around the articulatory target. Finally, the feedback may take the form of "target elevation," where the mean vector of the target distribution is shifted further away from the target value of the preceding phonological state compared with the situation when no Lombard effect occurs. Any of these three articulatory behavior changes may result in enhanced discriminability among speech units under severe environmental distortions, at the expense of greater articulatory efforts.

2.3.6   Piecewise Linearized Approximation for Articulatory-to-Acoustic Mapping

The nonlinearity $\mathbf{h}[\mathbf{z}(k)]$ of Eq. (2.6) is a source of difficulty in developing efficient and effective modeling learning algorithms. While the use of neural networks such as MLP or RBF as described in the preceding subsection makes it possible to design such algorithms, a more convenient strategy is to simplify the model by piecewise linearizing the nonlinear mapping function $\mathbf{h}[\mathbf{z}(k)]$ of Eq. (2.6). An extensive study based on nonlinear learning of the MLP-based model can be found in [24]), where a series of approximations are required to complete the algorithm development. Using piecewise linearization in the model would eliminate these approximations.

After the model simplification, it is hoped that the piecewise linear methods will lead to an adequate approximation to the nonlinear relationship between the hidden and observational spaces in the formulation of the dynamic speech model while gaining computational

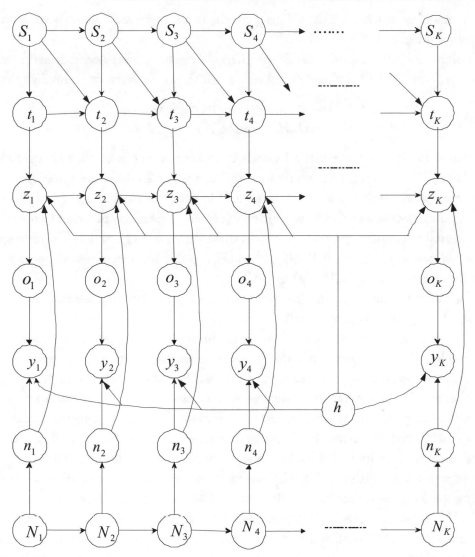

FIGURE 2.7: DBN that incorporates the Lombard effect in the comprehensive generative model of speech. The behavior of articulation is subject to modification (e.g., articulatory target overshoot or hyper-articulation or increased articulatory efforts by shortening time constant) under severe environmental distortions. This is represented by the "feedback" dependency from the noise nodes to the articulator nodes in the DBN

effectiveness and efficiency in model learning. The most straightforward method is to use a set of linear regression functions to replace the general nonlinear mapping in Eq. (2.6), while keeping intact the target-directed, linear state dynamics of Eq. (2.4). That is, rather than using one single set of linear-model parameters to characterize each phonological state, multiple sets of linear-model parameters are used. This gives rise to the mixture of linear dynamic model as extensively studied in [84, 85].

This piecewise linearized dynamic speech model can be written succinctly in the following state–space form (for a fixed phonological state s not shown for notational simplicity):

$$\mathbf{z}(k+1) = \Phi_m \mathbf{z}(k) + (\mathbf{I} - \Phi_m)\mathbf{t}_m + \mathbf{w}_m(k), \qquad (2.13)$$

$$\mathbf{o}(k) = \dot{\mathbf{H}}_m \dot{\mathbf{z}}(k) + \mathbf{v}_m(k), \quad m = 1, 2, \ldots, M, \qquad (2.14)$$

where $\dot{\mathbf{H}}_m = [\mathbf{a} \,|\, \mathbf{H}_m]$ is the expanded matrix by left appending vector $\mathbf{a}$ to matrix $\mathbf{H}_m$, and $\dot{\mathbf{z}}(k) = [1 \,|\, \mathbf{z}(k)']'$ is the expanded vector in a similar manner. In the above equations, M is the total number of mixture components in the model for each phonological state (e.g., phone). The state noise and measurement noise, $\mathbf{w}_m(k)$ and $\mathbf{v}_m(k)$, are respectively modeled by uncorrelated, IID, zero-mean, Gaussian processes with covariance matrices $\mathbf{Q}_m$ and $\mathbf{R}_m$. $\mathbf{o}$ represents the sequence of acoustic vectors, $\mathbf{o}(1), \mathbf{o}(2), \ldots, \mathbf{o}(k)\ldots$, and the $\mathbf{z}$ represents the sequence of hidden articulatory vectors, $\mathbf{z}(1), \mathbf{z}(2), \ldots, \mathbf{z}(k), \ldots$.

The full set of model parameters for each phonological state (not indexed for clarity) are $\Theta = (\Phi_m, \mathbf{t}_m, \mathbf{Q}_m, \mathbf{R}_m, \mathbf{H}_m, \text{ for } m = 1, 2, \ldots, M)$.

It is important to impose the following mixture-path constraint on the above dynamic system model: for each sequence of acoustic observations associated with a phonological state, the sequence is forced to be produced from a fixed mixture component, m, in the model. This means that the articulatory target for each phonological state is not permitted to switch from one mixture component to another within the duration of the same segment. The constraint is motivated by the physical nature of the dynamic speech model—the target that is correlated with its phonetic identity is defined at the segment level, not at the frame level. Use of the type of segment-level mixture is intended to represent the various sources of speech variability including speakers' vocal tract shape differences and speaking-habit differences, etc.

In Fig. 2.8 is shown the DBN representation for the piecewise linearized dynamic speech model as a simplified generative model of speech where the nonlinear mapping from hidden dynamic variables to acoustic observational variables is approximated by a piecewise linear relationship. The new, discrete random variable m is introduced to provide the "region" or mixture-component index m to the piecewise linear mapping. Both the input and output variables that are in a nonlinear relationship have now simultaneous dependency on m. The conditional PDFs involving this new node are

$$p[\mathbf{o}(k) \,|\, \mathbf{z}(k), m] = N[\mathbf{o}(k); \dot{\mathbf{H}}_m \dot{\mathbf{z}}(k), \mathbf{R}_m], \qquad (2.15)$$

and

$$p[\mathbf{z}(k+1) \,|\, \mathbf{z}(k), \mathbf{t}(k), \mathbf{s}, m] = N[\mathbf{z}(k+1); \Phi_{s,m}\mathbf{z}(k) - (\mathbf{I} - \Phi_{s,m})\mathbf{t}(k), \mathbf{Q}_m], \qquad (2.16)$$

where k denotes the time frame and s denotes the phonological state.

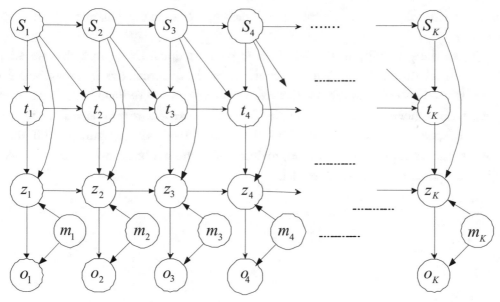

FIGURE 2.8: DBN representation for a mixture linear model as a simplified generative model of speech where the nonlinear mapping from hidden dynamic variables to acoustic observational variables is approximated by a piecewise linear relationship. The new, discrete random variable m is introduced to provide "region" index to the piecewise linear mapping. Both the input and output variables that are in a nonlinear relationship have now simultaneous dependency on m

2.4   SUMMARY

After providing general motivations and model design philosophy, technical detail of a multi-stage statistical generative model of speech dynamics and its associated computational framework based on DBN is presented in this chapter. We now summarize this model description. Equations (2.4) and (2.6) form a special version of the switching state–space model appropriate for describing the multilevel speech dynamics. The top-level dynamics occur at the discrete-state phonology, represented by the state transitions of **s** with a relatively long time scale (roughly about the duration of phones). The next level is the target (**t**) dynamics; it has the same time scale and provides systematic randomness at the segmental level. At the level of articulatory dynamics, the time scale is significantly shortened. This level represents the continuous-state dynamics driven by the stochastic target process as the input. The state equation (2.4) explicitly describes the dynamics in **z**, with index of s (which takes discrete values) implicitly representing the phonological process of transitions among a set of discrete states, which we call "switching." At the lowest level is acoustic dynamics, where there is no phonological switching process. Since the observation equation (2.6) is static, this simplified acoustic generation model assumes that acoustic dynamics are a direct consequence of articulatory dynamics only. Improvement of this model component that overcomes this simplification is unlikely until better modeling techniques are developed for representing multiple time scales in the dynamic aspects of speech acoustics.

Due to the generality of the DBN-based computational framework that we adopt, it becomes convenient to extend the above generative model of speech dynamics one step further from undistorted speech acoustics to distorted (or noisy) ones. We included this extension in this chapter. Another extension that includes the changed articulatory behavior due to acoustic distortion of speech is presented also within the same DBN-based computational framework. Finally, we discussed piecewise linear approximation in the nonlinear articulatory-to-acoustic mapping component of the overall model.

CHAPTER 3

Modeling: From Acoustic Dynamics to Hidden Dynamics

In Chapter 2, we described a rather general modeling scheme and the DBN-based computational framework for speech dynamics. Detailed implementation of the speech dynamic models would vary depending on the trade-offs in modeling precision and mathematical/algorithm tractability. In fact, various types of statistical models of speech beyond the HMM have already been in the literature for sometime, although most of them have not been viewed from a unified perspective as having varying degrees of approximation to the multistage speech chain. The purpose of this chapter is to take this unified view in classifying and reviewing a wide variety of current statistical speech models.

3.1 BACKGROUND AND INTRODUCTION

As we discussed earlier in this book, as a linguistic and physical abstraction, human speech production can be functionally represented at four distinctive but correlated levels of dynamics. The top level of the dynamics is symbolic or phonological. The multitiered linear sequence demonstrates the discrete, time-varying nature of speech dynamics at the mental motor-planning level of speech production. The next level of the dynamics is continuous-valued and associated with the functional, "task" variables in speech production. At this level, the goal or "task" of speech generation is defined, which may be either the acoustic goal such as vocal tract resonances or formants, or the articulatory goal such as vocal-tract constrictions, or their combination. It is at this level that each symbolic phonological unit is mapped to a unique set of the phonetic parameters. These parameter is often called the correlate of the phonological units. The third level of the dynamics occurs at the physiological articulators. Such articulatory dynamics are a nonlinear transformation of the task dynamics. Finally, the last level of the dynamics is the acoustic one, where speech "observations" are extracted from the speech signal. They are often called acoustic observations or "feature" vectors in automatic speech recognition applications, and are called speech measurements in experimental phonetics and speech science.

The review of several different types of computational dynamic models for speech in this chapter will be organized in view of the above functional levels of speech dynamics. We will classify the models into two main categories. In the first category are the models focusing on the lowest, acoustic level of dynamics, which is also the most peripheral level for human or computer speech perception. This class of models is often called the stochastic segment model as is known through an earlier review paper [14]. The second category consists of what is called the *hidden dynamic model* where the task dynamic and articulatory dynamic levels are functionally grouped into a functional single-level dynamics. In contrast to the acoustic-dynamic model, which represents coarticulation at the surface, observational level, the hidden dynamic model explores a deeper, unobserved (hence "hidden") level of the speech dynamic structure that regulates coarticulation and phonetic reduction.

3.2   STATISTICAL MODELS FOR ACOUSTIC SPEECH DYNAMICS

Hidden Markov model (HMM) is the simplest type of the acoustic dynamic model in this category. Stochastic segment models are a broad class of statistical models that generalize from the HMM and that intend to overcome some shortcomings of the HMM such as the conditional independent assumption and its consequences. As discussed earlier in this book, this assumption is grossly unrealistic and restricts the ability of the HMM as an accurate generative model. The generalization of the HMM by acoustic dynamic models is in the following sense: In an HMM, one frame of speech acoustics is generated by visiting each HMM state, while a variable-length sequence of speech frames is generated by visiting each "state" of a dynamic model. That is, a state in the acoustic dynamic or stochastic segment model is associated with a "segment" of acoustic speech vectors having a random sequence length.

Similar to an HMM, a stochastic segment model can be viewed as a generative process for observation sequences. It is intended to model the acoustic feature trajectories and temporal correlations that have been inadequately represented by an HMM. This is accomplished by introducing new parameters that characterize the trajectories and the temporal correlations.

From the perspective of the multilevel dynamics in the human speech process, the acoustic dynamic model can be viewed as a highly simplified model—collapsing all three lower phonetic levels of speech dynamics into one single level. As a result, the acoustic dynamic models have difficulties in capturing the structure of speech coarticulation and reduction. To achieve high performance in speech recognition, they tend to use many parallel (as opposed to hierarchical structured) parameters to model variability in acoustic dynamics, much like the strategies adopted by the HMM.

A convenient way to understand a variety of acoustic dynamic models and their relationships is to establish a hierarchy showing how the HMM is generalized by gradually relaxing the modeling assumptions. Starting with a conventional HMM in this hierarchy, there are two main classes of its extended or generalized models. Each of these classes further contains subclasses of models. We describe this hierarchy below.

3.2.1   Nonstationary-State HMMs

This model class has also been called the trended HMM, constrained mean trajectory model, segmental HMM, or stochastic trajectory model, etc., with minor variations according to whether the parameters defining the trend functions or trajectories are random or not and how their temporal properties are constrained. The trajectories for each state or segment are sometimes normalized in time, especially when the linguistic unit associated with the state is large (e.g., a word).

Given the HMM state s, the sample paths of most of these model types are explicitly defined acoustic feature trajectories:

$$\mathbf{o}(k) = \mathbf{g}_k(\mathbf{\Lambda}_s) + \mathbf{r}_s(k), \qquad (3.1)$$

where $\mathbf{g}_k(\mathbf{\Lambda}_s)$ is the deterministic function of time frame k, parameterized by state-specific $\mathbf{\Lambda}_s$, which can be either deterministic or random. And $\mathbf{r}_s(k)$ is a state-specific stationary residual signal.

The trend function $\mathbf{g}_k(\mathbf{\Lambda}_s)$ in Eq. (3.1) varies with time (as indexed by k), and hence describes acoustic dynamics. This is a special type of dynamics where no temporal recursion is involved in characterizing the time-varying function. Throughout this book, we call this special type of the dynamic function as a "trajectory," or a kinematic function.

We now discuss further classification of the nonstationary-state or trended HMMs.

Polynomial Trended HMM

In this subset of the nonstationary-state HMMs, the trend function associated with each HMM state is a polynomial function of time frames. Two common types of such models are as follows:

- *Observable polynomial trend functions*: This is the simplest trended HMM where there is no uncertainty in the polynomial coefficients $\mathbf{\Lambda}_s$ (e.g., [41,55,56,86]).

- *Random polynomial trend functions*: The trend functions $\mathbf{g}_k(\mathbf{\Lambda}_s)$ in Eq. (3.1) are stochastic due to the uncertainty in polynomial coefficients $\mathbf{\Lambda}_s$. $\mathbf{\Lambda}_s$ are random vectors in one of the two ways: (1) $\mathbf{\Lambda}_s$ has a discrete distribution [87,88] and (2) $\mathbf{\Lambda}_s$ has a continuous distribution. In the latter case, the model is called the segmental HMM, where the earlier versions have a polynomial order of zero [40,89] and the later versions have an order of one [90] or two [91].

Nonparametric Trended HMM

The trend function is determined by the training data after performing dynamic time warping [92], rather than by any parametric form.

Observation-dependent Trended Function

In this rather recent nonstationary-state HMM, the trend function is designed in a special way, where the parameters in $\mathbf{\Lambda}_s$ in the function $\mathbf{g}_k(\mathbf{\Lambda}_s)$ of Eq. (3.1) are made dependent on the observation vector $\mathbf{o}(k)$. The dependency is nonlinear, based on the posterior probability computation [26].

3.2.2 Multiregion Recursive Models

Common to this model class is the recursive form in dynamic modeling of the region-dependent time-varying acoustic feature vectors, where the "region" or state is often associated with a phonetic unit. The most typical recursion is of the following linear form:

$$\mathbf{o}(k) = \mathbf{\Lambda}_s(1)\mathbf{o}(k-1) + \cdots + \mathbf{\Lambda}_s(p)\mathbf{o}(k-p) + \mathbf{r}_s(k), \qquad (3.2)$$

and the starting point of the recursion for each state s comes usually from the previous state's ending history.

The model expressed in Eq. (3.2) provides clear contrast to the trajectory or trended models where the time-varying acoustic observation vectors are approximated as an explicit temporal function of time. The sample paths of the model Eq. (3.2), on the other hand, are piecewise, recursively defined stochastic time-varying functions. Further classification of this model class is discussed below.

Autoregressive or Linear-predictive HMM

In this model, the time-varying function associated with each region (a Markov state) is defined by linear prediction, or recursively defined autoregressive function. The work in [93] and that in [94] developed this type of model having the state-dependent linear prediction performed on the acoustic feature vectors (e.g., cepstra), with a first-order prediction and a second-order linear prediction, respectively. The work in [95, 96] developed the model having the state-dependent linear prediction performed on the speech waveforms. The latter model is also called the hidden filter model in [95].

Dynamics Defined by Jointly Optimized Static and Delta Parameters

In this more recently introduced HMM version with recursively defined state-bound dynamics on acoustic feature vectors, the dynamics are in the form of joint static and delta parameters [57, 97, 98]. The coefficients in the recursion are fixed for the delta parameters, instead of being optimized as in the linear-predictive HMM. The optimized feature-vector "trajectories" are obtained by joint use of static and delta model parameters. The results of the constrained optimization provide an explicit relationship between the static and delta acoustic features.

Nonlinear-predictive HMM

Several versions of nonlinear-predictive HMM have appeared in the literature, which generalize the linear prediction in Eq. (3.2) to nonlinear prediction using neural networks (e.g., [99–101]). In the model of [101], detailed statistical analysis was provided, proving that nonlinear prediction with a short temporal order effectively produces a correlation structure over a significantly longer temporal span.

Switching Linear Dynamic System Model

In this subclass of the multiregion recursive linear models, in addition to the use of the autoregressive function that recursively defines the region-bound dynamics, which we call (continuous)-state dynamics, a new noisy observation function is introduced. The actual effect of autoregression in this model is to smoothen the observed acoustic feature vectors. This model was originally introduced in [102] for speech modeling.

3.3   STATISTICAL MODELS FOR HIDDEN SPEECH DYNAMICS

The various types of acoustic dynamic or stochastic segment models described in this chapter generalize the HMM by generating a variable-length sequence of speech frames in each state, overcoming the HMM's assumption of local conditional independence. Yet the inconsistency between the HMM assumptions and the properties of the realistic dynamic speech process goes beyond this limitation. In acoustic dynamic models, the speech frames assigned to the same segment/state have been modeled to be temporally correlated and the model parameters been time-varying. However, the lengths of such segments are typically short. Longer-term correlation across phonetic units, which provides dynamic structure responsible for coarticulation and phonetic reduction, in a full utterance has not been captured.

This problem has been addressed by a class of more advanced dynamic speech models, which we call hidden dynamic models. A hidden dynamic model exploits an intermediate level of speech dynamics, functionally representing a combined system for speech motor control, task dynamics, and articulatory dynamics. This intermediate level is said to be hidden since it is not accessed directly from the speech acoustic data. The speech dynamics in this intermediate, hidden level explicitly captures the long-contextual-span properties over the phonetic units by imposing continuity constraints on the hidden dynamic variables internal to the acoustic observation data. The constraint is motivated by physical properties of speech generation. The constraint captures some key coarticulation and reduction properties in speech, and makes the model parameterization more parsimonious than does the acoustic dynamic model where modeling coarticulation requires a large number of free parameters. Since the underlying speech structure represented by the hidden dynamic model links a sequence of segments via continuity in the hidden dynamic variables, it can also be appropriately called the a super-segmental model.

Differing from the acoustic dynamic models, the hidden dynamic models represent speech structure by the hidden dynamic variables. Depending on the nature of these dynamic variables in light of multilevel speech dynamics discussed earlier, the hidden dynamic models can be broadly classified into

- articulatory dynamic model (e.g., [46, 54, 58, 59, 78, 79, 103, 104]);
- task-dynamic model (e.g., [105, 106]);
- vocal tract resonance (VTR) dynamic model (e.g., [24, 42, 48, 49, 84, 85, 107–112]);
- model with abstract dynamics (e.g., [42, 44, 107, 113]).

The VTR dynamics are a special type of task dynamics, with the acoustic goal or "task" of speech production in the VTR domain. Key advantages of using VTRs as the "task" are their direct correlation with the acoustic information, and the lower dimensionality in the VTR vector compared with the counterpart hidden vectors either in the articulatory dynamic model or in the task-dynamic model with articulatorily defined goal or "task" such as vocal tract constriction properties.

As an alternative classification scheme, the hidden dynamic models can also be classified, from the computational perspective, according to whether the hidden dynamics are represented mathematically with temporal recursion or not. Like the acoustic dynamic models, the two types of the hidden dynamic models in this classification scheme are reviewed here.

3.3.1 Multiregion Nonlinear Dynamic System Models

The hidden dynamic models in this first model class use the temporal recursion (k-recursion via the predictive function $\mathbf{g}_k$ in Eq. (3.3)) to define the hidden dynamics $\mathbf{z}(k)$. Each region, s, of such dynamics is characterized by the s-dependent parameter set $\mathbf{\Lambda}_s$, with the "state noise" denoted by $\mathbf{w}_s(k)$. The memoryless nonlinear mapping function is exploited to link the hidden dynamic vector $\mathbf{z}(k)$ to the observed acoustic feature vector $\mathbf{o}(k)$, with the "observation noise" denoted by $\mathbf{v}_s(k)$, and also parameterized by region-dependent parameters. The combined "state equation" (3.3) and "observation equation" (3.4) form a general multiregion nonlinear dynamic system model:

$$\mathbf{z}(k+1) = \mathbf{g}_k[\mathbf{z}(k), \mathbf{\Lambda}_s] + \mathbf{w}_s(k), \tag{3.3}$$

$$\mathbf{o}(k') = \mathbf{h}_{k'}[\mathbf{z}(k'), \mathbf{\Omega}_{s'}] + \mathbf{v}_{s'}(k'). \tag{3.4}$$

where subscripts k and k' indicate that the functions $\mathbf{g}[\cdot]$ and $\mathbf{h}[\cdot]$ are time-varying and may be asynchronous with each other. The subscripts s or s' denotes the dynamic region correlated with phonetic categories.

Various simplified implementations of the above generic nonlinear system model have appeared in the literature (e.g., [24, 33, 42, 45, 46, 59, 85, 108]). Most of these implementations reduce the predictive function $\mathbf{g}_k$ in the state equation (3.3) into a linear form, and use the concept of phonetic targets as part of the parameters. This gives rise to linear target filtering (by infinite impulse response or IIR filters) as a model for the hidden dynamics. Also, many of these implementations use neural networks as the nonlinear mapping function $\mathbf{h}_k[\mathbf{z}(k), \mathbf{\Omega}_s]$ in the observation equation (3.4).

3.3.2   Hidden Trajectory Models

The second type of the hidden dynamic models use trajectories (i.e., explicit functions of time with no recursion) to represent the temporal evolution of the hidden dynamic variables (e.g., VTR or articulatory vectors). This *hidden* trajectory model (HTM) differs conceptually from the acoustic dynamic or trajectory model in that the articulatory-like constraints and structure are captured in the HTM via the continuous-valued hidden variables that run across the phonetic units. Importantly, the polynomial trajectories, which were shown to fit well to the temporal properties of cepstral features [55, 56], are not appropriate for the hidden dynamics that require realistic physical constraints of segment-bound monotonicity and target-directedness. One parametric form of the hidden trajectory constructed to satisfy both these constraints is the critically damped exponential function of time [33, 114]. Another parametric form of the hidden trajectory, which also satisfies these constraints but with more flexibility to handle asynchrony between segment boundaries for the hidden trajectories and for the acoustic features, has been developed more recently [109, 112, 115, 116] based on finite impulse response (FIR) filtering of VTR target sequences. In Chapter 5, we provide a systematic account of this model, synthesizing and expanding the earlier descriptions of this work in [109, 115, 116].

3.4   SUMMARY

This chapter serves as a bridge between the general modeling and computational framework for speech dynamics (Chapter 2) and Chapters 4 and 5 on detailed descriptions of two specific implementation strategies and algorithms for hidden dynamic models. The theme of this chapter is to move from the relatively simplistic view of dynamic speech modeling confined within the acoustic stage to the more realistic view of multistage speech dynamics with an intermediate hidden dynamic layer between the phonological states and the acoustic dynamics. The latter, with appropriate constraints in the form of the dynamic function, permits a representation of the underlying speech structure responsible for coarticulation and speaking-effort-related reduction. This type of structured modeling is difficult to accomplish by acoustic dynamic models with no hidden dynamic layer, unless highly elaborate model parameterization is carried out. In Chapter 5, we will show an example where a hidden trajectory model can be simplified to an equivalent of an acoustic trajectory model whose trajectory parameters become long-span context-dependent via a structured means and delicate parameterization derived from the construction of the hidden trajectories.

Guided by this theme, in this chapter we classify and review a rather rich body of literature on a wide variety of statistical models of speech, starting with the traditional HMM [4] as the most primitive model. Two major classes of the models, acoustic dynamic models and hidden dynamic models, respectively, are each further classified into subclasses based on how the dynamic functions are constructed. When explicit temporal functions are constructed without recursion, then we have classes of "trajectory" models. The trajectory models and recursively defined dynamic models can achieve a similar level of modeling accuracy but they demand very different algorithm development for model parameter learning and for speech decoding. Each of these two classes (acoustic vs. hidden dynamic) and two types (trajectory vs. recursive) of the models simplifies, in different ways, the DBN structure as the general computational framework for the full multistage speech chain (Chapter 2).

In the remaining two chapters, we select two types of hidden dynamic models of speech for their detailed exposition, one with and another without recursion in defining the hidden dynamic variables. The exposition will include the implementation strategies (discretization of the hidden dynamic variables or otherwise) and the related algorithms for model parameter learning and model scoring/decoding. The implementation strategy with discretization of recursively defined hidden speech dynamics will be covered in Chapter 4, and the strategy using hidden trajectories (i.e., explicit temporal functions) with no discretization will be discussed in Chapter 5.

CHAPTER 4

Models with Discrete-Valued Hidden Speech Dynamics

In this chapter, we focus on a special type of hidden dynamic models where the hidden dynamics are recursively defined and where these hidden dynamic values are discretized. The discretization or quantization of the hidden dynamics causes an approximation to the original continuous-valued dynamics as described in the earlier chapters but it enables an implementation strategy that can take direct advantage of the forward–backward algorithm and dynamic programming in model parameter learning and decoding. Without discretization, the parameter learning and decoding problems would be typically intractable (i.e., the computation cost would increase exponentially with time). Under different kinds of model implmentation schemes, other types of approximation will be needed and one type of the approximation in this case will be detailed in Chapter 5.

This chapter is based on the materials published in [110, 117], with reorganization, rewriting, and expansion of these materials so that they naturally fit as an integral part of this book.

4.1   BASIC MODEL WITH DISCRETIZED HIDDEN DYNAMICS

In the basic model presented in this section, we assume discrete-time, first-order hidden dynamics in the state equation and linearized mapping from the hidden dynamic variables to the acoustic observation variables in the observation equation. Before discretizing hidden dynamics, the first-order dynamics in a scalar form have the following form (which was discussed in Chapter 2 with a vector form):

$$x_t = r_s x_{t-1} + (1 - r_s) T_s + w_t(s), \tag{4.1}$$

where state noise $w_t \sim N(w_k; 0, B_s)$ is assumed to be IID, zero-mean Gaussian with phonological state (s)-dependent precision (inverse of variance) B_s. The linearized observation equation is

$$o_t = H_s x_t + h_s + v_t, \tag{4.2}$$

where observation noise $v_k \sim N(v_k; 0, D_s)$ is assumed to be IID, zero-mean Gaussian with precision D_s.

We now perform discretization or quantization on hidden dynamic variable x_t. For simplicity in illustration, we use scalar hidden dynamics most of the times in this chapter (except Section 4.2.3) where scalar quantization is carried out, and let C denote the total number of discretization/quantization levels. (For the more realistic, multidimensional hidden dynamic case, C would be the total number of cells in the vector-quantized space.) In the following derivation of the EM algorithm for parameter learning, we will use variable $x_t[i]$ or i_t to denote the event that at time frame t the state variable (or vector) x_t takes the mid-point (or centroid) value associated with the ith discretization level in the quantized space.

We now describe this basic model with discretized hidden dynamics in an explicit probabilistic form and then derive and present a maximum-likelihood (ML) parameter estimation technique based on the Expectation-Maximization (EM) algorithm. The background information on ML and EM can be found in of [9], [Part I, Ch. 5, Sec. 5.6].

4.1.1   Probabilistic Formulation of the Basic Model

Before discretization, the basic model that consists of Eqs. (4.1) and (4.2) can be equivalently written in the following explicit probabilistic form:

$$p(x_t \mid x_{t-1}, s_t = s) = N(x_t; r_s x_{t-1} + (1 - r_s)T_s, B_s), \tag{4.3}$$

$$p(o_t \mid x_t, s_t = s) = N(o_t; H_s x_t + h_s, D_s). \tag{4.4}$$

And we also have the transition probability for the phonological states:

$$p(s_t = s \mid s_{t-1} = s') = \pi_{s's}.$$

Then the joint probability can be written as

$$p(s_1^N, x_1^N, o_1^N) = \prod_{t=1}^{N} \pi_{s_{t-1}s_t} p(x_t \mid x_{t-1}, s_t) p(o_t \mid x_t, s_t = s),$$

where N is the total number of observation data points in the training set.

After discretization of hidden dynamic variables, Eqs. (4.3) and (4.4) are approximated as

$$p(x_t[i] \mid x_{t-1}[j], s_t = s) \approx N(x_t[i]; r_s x_{t-1}[j] + (1 - r_s)T_s, B_s), \tag{4.5}$$

and

$$p(o_t \mid x_t[i], s_t = s) \approx N(o_t; H_s x_t[i] + h_s, D_s). \tag{4.6}$$

4.1.2   Parameter Estimation for the Basic Model: Overview

For carrying out the EM algorithm for parameter estimation of the above discretized model, we first establish the auxiliary function, Q. Then we simplify the Q function into a form that can be optimized in a closed form.

According to the EM theory, the auxiliary objective function Q is the conditional expectation of logarithm of the joint likelihood of all hidden and observable variables. The conditioning events are all observation sequences in the training data:

$$o_1^N = o_1, o_2, \ldots, o_t, \ldots, o_N,$$

And the expectation is taken over the posterior probability for all hidden variable sequences:

$$x_1^N = x_1, x_2, \ldots, x_t, \ldots, x_N,$$

and

$$s_1^N = s_1, s_2, \ldots, s_t, \ldots, s_N,$$

This gives (before discretization of the hidden dynamic variables):

$$Q = \sum_{s_1} \cdots \sum_{s_t} \cdots \sum_{s_N} \int_{x_1} \cdots \int_{x_t} \cdots \int_{x_N} p(s_1^N, x_1^N \mid o_1^N) \log p(s_1^N, x_1^N, o_1^N) dx_1 \cdots dx_t \cdots dx_N,$$

$$(4.7)$$

where the summation for each phonological state s is from 1 to S (the total number of distinct phonological units).

After discretizing x_t into $x_t[i]$, the objective function of Eq. (4.7) is approximated by

$$Q \approx \sum_{s_1} \cdots \sum_{s_t} \cdots \sum_{s_N} \sum_{i_1} \cdots \sum_{i_t} \cdots \sum_{i_N} p(s_1^N, i_1^N \mid o_1^N) \log p(s_1^N, i_1^N, o_1^N), \qquad (4.8)$$

where the summation for each discretization index i is from 1 to C.

We now describe details of the E-step and m-Step in the EM algorithm.

4.1.3   EM Algorithm: The E-Step

The following outlines the simplification steps for the objective function of Eq. (4.8). Let us denote the sequence summation $\sum_{s_1} \cdots \sum_{s_t} \cdots \sum_{s_N}$ by $\sum_{s_1^N}$, and summation $\sum_{i_1} \cdots \sum_{i_t} \cdots \sum_{i_N}$ by $\sum_{i_1^N}$. Then we rewrite Q in Eq. (4.8) as

$$Q(r_s, T_s, B_s, H_s, h_s, D_s) \approx \sum_{s_1^N} \sum_{i_1^N} p(s_1^N, i_1^N \mid o_1^N) \log p(s_1^N, i_1^N, o_1^N) \qquad (4.9)$$

$$= \underbrace{\sum_{s_1^N} \sum_{i_1^N} p(s_1^N, i_1^N \mid o_1^N) \log p(o_1^N \mid s_1^N, i_1^N)}_{Q_o(H_s, h_s, D_s)} + \underbrace{\sum_{s_1^N} \sum_{i_1^N} p(s_1^N, i_1^N \mid o_1^N) \log p(s_1^N, i_1^N)}_{Q_x(r_s, T_s, B_s)},$$

where

$$p(s_1^N, i_1^N) = \prod_t \pi_{s_{t-1}s_t} N(x_t[i]; r_{s_t} x_{t-1}[j] + (1 - r_{s_t}) T_{s_t}, B_{s_t}),$$

and

$$p(o_1^N \mid s_1^N, i_1^N) = \prod_t N(o_t; H_{s_t} x_t[i] + h_{s_t}, D_{s_t}).$$

In these equations, discretization indices i and j denote the hidden dynamic values taken at time frames t and $t - 1$, respectively. That is, $s_t = i$, $s_{t-1} = j$.

We first compute Q_o (omitting constant $-0.5d \, \log(2\pi)$ that is irrelevant to optimization):

$$Q_o = 0.5 \sum_{s_1^N} \sum_{i_1^N} p(s_1^N, i_1^N \mid o_1^N) \sum_{t=1}^{N} \left[\log|D_{s_t}| - D_{s_t} \left(o_t - H_{s_t} x_t[i] - h_{s_t} \right)^2 \right]$$

$$= \sum_{s=1}^{S} \sum_{i=1}^{C} \left\{ 0.5 \sum_{s_1^N} \sum_{i_1^N} p(s_1^N, i_1^N \mid o_1^N) \sum_{t=1}^{N} \left[\log|D_{s_t}| - D_{s_t} \left(o_t - H_{s_t} x_t[i] - h_{s_t} \right)^2 \right] \right\} \delta_{s_t s} \delta_{i_t i}$$

$$= 0.5 \sum_{s=1}^{S} \sum_{i=1}^{C} \sum_{t=1}^{N} \left\{ \sum_{s_1^N} \sum_{i_1^N} p(s_1^N, i_1^N \mid o_1^N) \delta_{s_t s} \delta_{i_t i} \right\} \left[\log|D_s| - D_s \left(o_t - H_s x_t[i] - h_s \right)^2 \right].$$

Noting that

$$\sum_{s_1^N} \sum_{i_1^N} p(s_1^N, i_1^N \mid o_1^N) \delta_{s_t s} \delta_{i_t i} = p(s_t = s, i_t = i \mid o_1^N) = \gamma_t(s, i),$$

we obtain the simplified form of

$$Q_o(H_s, h_s, D_s) = 0.5 \sum_{s=1}^{S} \sum_{t=1}^{N} \sum_{i=1}^{C} \gamma_t(s, i) \left[\log|D_s| - D_s \left(o_t - H_s x_t[i] - h_s \right)^2 \right]. \qquad (4.10)$$

Similarly, after omitting optimization-independent constants, we have

$$Q_x = 0.5 \sum_{s_1^N} \sum_{i_1^N} p(s_1^N, i_1^N \mid o_1^N) \sum_{t=1}^{N} \left[\log|B_{s_t}| - B_{s_t} \left(x_t[i] - r_{s_t} x_{t-1}[j] - (1 - r_{s_t}) T_{s_t} \right)^2 \right]$$

$$= \sum_{s=1}^{S} \sum_{i=1}^{C} \sum_{j=1}^{C} \left\{ 0.5 \sum_{s_1^N} \sum_{i_1^N} p(s_1^N, i_1^N \mid o_1^N) \right.$$

$$\times \left. \sum_{t=1}^{N} \left[\log|B_{s_t}| - B_{s_t} \left(x_t[i] - r_{s_t} x_{t-1}[j] - (1 - r_{s_t}) T_{s_t} \right)^2 \right] \right\} \delta_{s_t s} \delta_{i_t i} \delta_{i_{t-1} j}$$

$$= 0.5 \sum_{s=1}^{S} \sum_{i=1}^{C} \sum_{j=1}^{C} \sum_{t=1}^{N} \left\{ \sum_{s_1^N} \sum_{i_1^N} p(s_1^N, i_1^N \mid o_1^N) \delta_{s_t s} \delta_{i_t i} \delta_{i_{t-1} j} \right\}$$

$$\times \left[\log |B_s| - B_s \left(x_t[i] - r_s x_{t-1}[j] - (1 - r_s) T_s \right)^2 \right].$$

Now noting

$$\sum_{s_1^N} \sum_{i_1^N} p(s_1^N, i_1^N \mid o_1^N) \delta_{s_t s} \delta_{i_t i} \delta_{i_{t-1} j} = p(s_t = s, i_t = i, i_{t-1} = j \mid o_1^N) = \xi_t(s, i, j),$$

we obtain the simplified form of

$$Q_x(r_s, T_s, B_s) = 0.5 \sum_{s=1}^{S} \sum_{t=1}^{N} \sum_{i=1}^{C} \sum_{j=1}^{C} \xi_t(s, i, j) \Big[\log |B_s|$$

$$- B_s \left(x_t[i] - r_s x_{t-1}[j] - (1 - r_s) T_s \right)^2 \Big], \qquad (4.11)$$

Note that large computational saving can be achieved by limiting the summations in Eq. (4.11) for i, j based on the relative smoothness of hidden dynamics. That is, the range of i, j can be limited such that $|x_t[i] - x_{t-1}[j]| < \text{Th}$, where Th is empirically set threshold value that controls the computation cost and accuracy.

In Eqs. (4.11) and (4.10), we used $\xi_t(s, i, j)$ and $\gamma_t(s, i)$ to denote the single-frame posteriors of

$$\xi_t(s, i, j) \equiv p(s_t = s, x_t[i], x_{t-1}[j] \mid o_1^N),$$

and

$$\gamma_t(s, i) \equiv p(s_t = s, x_t[i] \mid o_1^N).$$

These can be computed efficiently using the generalized forward–backward algorithm (part of the E-step), which we describe below.

4.1.4 A Generalized Forward–Backward Algorithm

The only quantities that need to be determined in simplified auxiliary function $Q = Q_o + Q_x$ as in Eqs. (4.9)–(4.11) are the two frame-level posteriors $\xi_t(s, i, j)$ and $\gamma_t(s, i)$, which we compute now in order to complete the E-step in the EM algorithm.

Generalized $\alpha(s_t, i_t)$ Forward Recursion

The generalized forward recursion discussed here uses a new definition of the variable

$$\alpha_t(s, i) = p(o_1^t, s_t = s, i_t = i).$$

The generalization of the standard forward–backward algorithm for HMM in any standard textbook on speech recognition is by including additional discrete hidden variables related to hidden dynamics.

For notational convenience, we use $\alpha(s_t, i_t)$ to denote $\alpha_t(s, i)$ below. The forward recursive formula is

$$\alpha(s_{t+1}, i_{t+1}) = \sum_{s_t=1}^{S} \sum_{i_t=1}^{C} \alpha(s_t, i_t) p(s_{t+1}, i_{t+1}|s_t, i_t) p(o_{t+1}|s_{t+1}, i_{t+1}). \tag{4.12}$$

Proof of Eq. (4.12):

$$\alpha(s_{t+1}, i_{t+1}) \equiv p(o_1^{t+1}, s_{t+1}, i_{t+1})$$

$$= \sum_{s_t} \sum_{i_t} p(o_1^t, o_{t+1}, s_{t+1}, i_{t+1}, s_t, i_t)$$

$$= \sum_{s_t} \sum_{i_t} p(o_{t+1}, s_{t+1}, i_{t+1} \mid o_1^t, s_t, i_t) p(o_1^t, s_t, i_t)$$

$$= \sum_{s_t} \sum_{i_t} p(o_{t+1}, s_{t+1}, i_{t+1} \mid s_t, i_t) \alpha(s_t, i_t)$$

$$= \sum_{s_t} \sum_{i_t} p(o_{t+1} \mid s_{t+1}, i_{t+1}, s_t, i_t) p(s_{t+1}, i_{t+1} \mid s_t, i_t) \alpha(s_t, i_t)$$

$$= \sum_{s_t} \sum_{i_t} p(o_{t+1} \mid s_{t+1}, i_{t+1}) p(s_{t+1}, i_{t+1} \mid s_t, i_t) \alpha(s_t, i_t). \tag{4.13}$$

In Eq. (4.12), $p(o_{t+1} \mid s_{t+1}, i_{t+1})$ is determined by the observation equation:

$$p(o_{t+1} \mid s_{t+1} = s, i_{t+1} = i) = N(o_{t+1}; H_s x_{t+1}[i] + h_s, D_s),$$

and $p(s_{t+1}, i_{t+1} \mid s_t, i_t)$ is determined by the state equation (with order one) and the switching Markov chain's transition probabilities:

$$p(s_{t+1} = s, i_{t+1} = i \mid s_t = s', i_t = i') \approx p(s_{t+1} = s \mid s_t = s') p(i_{t+1} = i \mid i_t = i')$$
$$= \pi_{s_{t-1} s_t} p(i_{t+1} = i \mid i_t = i'). \tag{4.14}$$

Generalized $\gamma(s_t, i_t)$ Backward Recursion

Rather than performing backward β recursion and then combining the α and β to obtain the single-frame posterior as for the conventional HMM, a more memory-efficient technique can be used for backward recursion, which directly computes the single-frame posterior. For notational convenience, we use $\gamma(s_t, i_t)$ to denote $\gamma_t(s, i)$ below.

The development of the generalized $\gamma(s_t, i_t)$ backward recursion for the first-order state equation proceeds as follows:

$$\gamma(s_t, i_t) \equiv p(s_t, i_t \mid o_1^N)$$

$$= \sum_{s_{t+1}} \sum_{i_{t+1}} p(s_t, i_t, s_{t+1}, i_{t+1} \mid o_1^N)$$

$$= \sum_{s_{t+1}} \sum_{i_{t+1}} p(s_t, i_t, s_{t+1}, i_{t+1} \mid o_1^N) p(s_{t+1}, i_{t+1} \mid o_1^N)$$

$$= \sum_{s_{t+1}} \sum_{i_{t+1}} p(s_t, i_t, s_{t+1}, i_{t+1} \mid o_1^t) \gamma(s_{t+1}, i_{t+1})$$

$$= \sum_{s_{t+1}} \sum_{i_{t+1}} \frac{p(s_t, i_t, s_{t+1}, i_{t+1}, o_1^t)}{p(s_{t+1}, i_{t+1}, o_1^t)} \gamma(s_{t+1}, i_{t+1}) \qquad \text{(Bayes rule)}$$

$$= \sum_{s_{t+1}} \sum_{i_{t+1}} \frac{p(s_t, i_t, s_{t+1}, i_{t+1}, o_1^t)}{\sum_{s_t} \sum_{i_t} p(s_t, i_t, s_{t+1}, i_{t+1}, o_1^t)} \gamma(s_{t+1}, i_{t+1})$$

$$= \sum_{s_{t+1}} \sum_{i_{t+1}} \frac{p(s_t, i_t, o_1^t) p(s_{t+1}, i_{t+1} \mid s_t, i_t, o_1^t)}{\sum_{s_t} \sum_{i_t} p(s_t, i_t, o_1^t) p(s_{t+1}, i_{t+1} \mid s_t, i_t, o_1^t)} \gamma(s_{t+1}, i_{t+1})$$

$$= \sum_{s_{t+1}} \sum_{i_{t+1}} \frac{\alpha(s_t, i_t) p(s_{t+1}, i_{t+1} \mid s_t, i_t)}{\sum_{s_t} \sum_{i_t} \alpha(s_t, i_t) p(s_{t+1}, i_{t+1} \mid s_t, i_t)} \gamma(s_{t+1}, i_{t+1}), \qquad (4.15)$$

where the last step uses conditional independence, and where $\alpha(s_t, i_t)$ and $p(s_{t+1}, i_{t+1} \mid s_t, i_t)$ on the right-hand side of Eq. (4.15) have been computed already in the forward recursion. Initialization for the above γ recursion is $\gamma(s_N, i_N) = \alpha(s_N, i_N)$, which will be equal to 1 for the left-to-right model of phonetic strings.

Given this result, $\xi_t(s, i, j)$ can be computed directly using $\alpha(s_t, i_t)$ and $\gamma(s_t, i_t)$. Both of them are already computed from the forward–backward recursions described above.

Alternatively, we can compute β generalized recursion (not discussed here) and then combine αs and βs to obtain $\gamma_t(s, i)$ and $\xi_t(s, i, j)$.

4.1.5   EM Algorithm: The M-Step

Given the results of the E-step described above where the frame-level posteriors are computed efficiently by the generalized forward–backward algorithm, we now derive the reestimation formulas, as the M-step in the EM algorithm, by optimizing the simplified auxiliary function $Q = Q_o + Q_x$ as in Eqs. (4.9), (4.10) and (4.11).

Reestimation for the Hidden-to-Observation Mapping Parameters H_s and h_s

Taking partial derivatives of Q_o in Eq. (4.10) with respect to H_s and h_s, respectively, and setting them to zero, we obtain:

$$\frac{\partial Q_o(H_s, h_s, D_s)}{\partial h_s} = -D_s \sum_{t=1}^{N} \sum_{i=1}^{C} \gamma_t(s, i)\{o_t - H_s x_t[i] - h_s\} = 0, \tag{4.16}$$

and

$$\frac{\partial Q_o(H_s, h_s, D_s)}{\partial H_s} = -D_s \sum_{t=1}^{N} \sum_{i=1}^{C} \gamma_t(s, i)\{o_t - H_s x_t[i] - h_s\}x_t[i] = 0. \tag{4.17}$$

These can be rewritten as the standard linear system of equations:

$$U\hat{H}_s + V_1 \hat{h}_s = C_1, \tag{4.18}$$

$$V_2 \hat{H}_s + U\hat{h}_s = C_2, \tag{4.19}$$

where

$$U = \sum_{t=1}^{N} \sum_{i=1}^{C} \gamma_t(s, i)x_t[i], \tag{4.20}$$

$$V_1 = N, \tag{4.21}$$

$$C_1 = \sum_{t=1}^{N} \sum_{i=1}^{C} \gamma_t(s, i)o_t, \tag{4.22}$$

$$V_2 = \sum_{t=1}^{N} \sum_{i=1}^{C} \gamma_t(s, i)x_t^2[i], \tag{4.23}$$

$$C_2 = \sum_{t=1}^{N} \sum_{i=1}^{C} \gamma_t(s, i)o_t x_t[i]. \tag{4.24}$$

The solution is

$$\begin{bmatrix} \hat{H}_s \\ \hat{h}_s \end{bmatrix} = \begin{bmatrix} U & V_1 \\ V_2 & U \end{bmatrix}^{-1} \begin{bmatrix} C_1 \\ C_2 \end{bmatrix}. \tag{4.25}$$

Reestimation for the Hidden Dynamic Shaping Parameter r_s

Taking partial derivative of Q_x in Eq. (4.11) with respect to r_s and setting it to zero, we obtain the reestimation formula of

$$\frac{\partial Q_x(r_s, T_s, B_s)}{\partial r_s} = -B_s \sum_{t=1}^{N} \sum_{i=1}^{C} \sum_{j=1}^{C} \xi_t(s, i, j) \tag{4.26}$$

$$\times \left[x_t[i] - r_s x_{t-1}[j] - (1 - r_s)T_s \right]\left[x_{t-1}[j] - T_s \right] = 0.$$

Solving for r_s, we have

$$
\hat{r}_s = \left[\sum_{t=1}^{N} \sum_{i=1}^{C} \sum_{j=1}^{C} \xi_t(s, i, j)(T_s - x_{t-1}[j])^2 \right]^{-1}
$$
$$
\times \left[\sum_{t=1}^{N} \sum_{i=1}^{C} \sum_{j=1}^{C} \xi_t(s, i, j)(T_s - x_{t-1}[j])(T_s - x_t[i]) \right], \tag{4.27}
$$

where we assume that all other model parameters are fixed.

It is interesting to note from the above that when x_t is monotonically moving (on average) toward the target T_s (i.e., no target overshooting), the reestimate of r_s is guaranteed to be positive, as it should be.

Reestimation for the Hidden Dynamic Target Parameter T_s

Similarly, taking partial derivative of Q_x in Eq. (4.11) with respect to T_s and setting it to zero, we obtain the reestimation formula of

$$
\frac{\partial Q_x(r_s, T_s, B_s)}{\partial T_s} = -B_s \sum_{t=1}^{N} \sum_{i=1}^{C} \sum_{j=1}^{C} \xi_t(s, i, j)\left[x_t[i] - r_s x_{t-1}[j] - (1 - r_s)T_s \right](1 - r_s) = 0. \tag{4.28}
$$

Solving for T_s, we obtain

$$
\hat{T}_s = \frac{1}{1 - r_s} \sum_{t=1}^{N} \sum_{i=1}^{C} \sum_{j=1}^{C} \xi_t(s, i, j)\left[x_t[i] - r_s x_{t-1}[j] \right]. \tag{4.29}
$$

Intuitions behind the target estimate above are particularly obvious.

Reestimation for the Noise Precisions B_s and D_s

Setting

$$
\frac{\partial Q_x(r_s, T_s, B_s)}{\partial B_s} = 0.5 \sum_{t=1}^{N} \sum_{i=1}^{C} \sum_{j=1}^{C} \xi_t(s, i, j)\left[B_s^{-1} - (x_t[i] - r_s x_{t-1}[j] - (1 - r_s)T_s)^2 \right] = 0,
$$

we obtain the state noise variance reestimate of

$$
\hat{B}_s^{-1} = \frac{\sum_{t=1}^{N} \sum_{i=1}^{C} \sum_{j=1}^{C} \xi_t(s, i, j)\left[x_t[i] - \hat{r}_s x_{t-1}[j] - (1 - \hat{r}_s)T_s \right]^2}{\sum_{t=1}^{N} \sum_{i=1}^{C} \sum_{j=1}^{C} \xi_t(s, i, j)}. \tag{4.30}
$$

Similarly, setting

$$
\frac{\partial Q_o(H_s, h_s, D_s)}{\partial D_s} = 0.5 \sum_{t=1}^{N} \sum_{i=1}^{C} \gamma_t(s, i)\left[D_s^{-1} - (o_t - H_s x_t[i] - h_s)^2 \right] = 0,
$$

we obtain the observation noise variance reestimate of

$$\hat{D}_s = \frac{\sum_{t=1}^{N} \sum_{i=1}^{C} \gamma_t(s, i) [o_t - H_s x_t[i] - h_s]^2}{\sum_{t=1}^{N} \sum_{i=1}^{C} \gamma_t(s, i)}.$$ (4.31)

4.1.6   Decoding of Discrete States by Dynamic Programming

After the parameters of the basic model are estimated using the EM algorithm described above, estimation of discrete phonological states and of the quantized hidden dynamic variables can be carried out jointly. We call this process "decoding." Estimation of the phonological states is the problem of speech recognition, and estimation of the hidden dynamic variables is the problem of tracking hidden dynamics. For large vocabulary speech recognition, aggressive pruning and careful design of data structures will be required (which is not described in this book).

Before describing the decoding algorithm, which is aimed at finding the best single joint state and quantized hidden dynamic variable sequences (s_1^N, i_1^N) for a given observation sequence o_1^N, let us define the quantity

$$\delta_t(s, i) = \max_{s_1, s_2, \dots, s_{t-1}, i_1, i_2, \dots, i_{t-1}} P(o_1^t, s_1^{t-1}, i_1^{t-1}, s_t = s, x_t[i])$$

$$= \max_{s_1^{t-1}, i_1^{t-1}} P(o_1^t, s_1^{t-1}, i_1^{t-1}, s_t = s, i_t = i).$$ (4.32)

Note that each $\delta_t(s, i)$ defined here is associated with a node in a three-dimensional trellis diagram. Each increment in time corresponds to reaching a new stage in dynamic programming (DP). At the final stage $t = N$, we have the objective function of $\delta_N(s, i)$ that is accomplished via all the previous stages of computation for $t \leq N - 1$. On the basis of the DP optimality principle, the optimal (joint) partial likelihood at the processing stage of $t + 1$ can be computed using the following DP recursion:

$$\delta_{t+1}(s, i) = \max_{s', i'} \delta_t(s', i') p(s_{t+1} = s, i_{t+1} = i \mid s_t = s', i_t = i') p(o_{t+1} \mid s_{t+1} = s, i_{t+1} = i)$$

$$\approx \max_{s', i'} \delta_t(s', i') p(s_{t+1} = s \mid s_t = s') p(i_{t+1} = i \mid i_t = i') p(o_{t+1} \mid s_{t+1} = s, i_{t+1} = i)$$

$$= \max_{s', i', j} \delta_t(s', i') \pi_{s's} N(x_{t+1}[i]; r_{s'} x_t[j] + (1 - r_{s'}) T_{s'}, B_{s'})$$

$$\times N(o_{t+1}; H_s x_{t+1}[i] + h_s, D_s),$$ (4.33)

for all states s and for all quantization indices i. Each pair of (s, i) at this processing stage is a hypothesized "precursor" node in the global optimal path. All such nodes except one will be eventually eliminated after the backtracking operation. The essence of DP used here is that we only need to compute the quantities of $\delta_{t+1}(s, i)$ as individual nodes in the trellis, removing the need to keep track of a very large number of partial paths from the initial stage to the current $(t + 1)th$ stage, which would be required for the exhaustive search. The optimality is guaranteed, due to the DP optimality principle, with the computation only linearly, rather than geometrically, increasing with the length N of the observation data sequence o_1^N.

4.2   EXTENSION OF THE BASIC MODEL

The preceding section presented details of the basic hidden dynamic model where the discretized state equation takes the simplest first-order recursive form and where the observation equation also takes the simplest linear form responsible for mapping from hidden dynamic variables to acoustic observation variables. We now present an extension of this discretized basic model. First, we will extend the state equation of the basic model from first-order dynamics to second-order dynamics so as to improve the modeling accuracy. Second, we will extend the observation equation of the basic model from the linear form to a nonlinear form of the mapping function from the discretized hidden dynamic variables to (nondiscretized or continuous-valued) acoustic observation variables.

4.2.1   Extension from First-Order to Second-Order Dynamics

In this first step of extension of the basic model, we change from the first-order state equation (Eq. (4.1)):

$$x_t = r_s x_{t-1} + (1 - r_s) T_s + w_t(s),$$

to the new second-order state equation

$$x_t = 2 r_s x_{t-1} - r_s^2 x_{t-2} + (1 - r_s)^2 T_s + w_t(s). \tag{4.34}$$

Here, like the first-order state equation, state noise $w_k \sim N(w_k; 0, B_s)$ is assumed to be IID zero-mean Gaussian with state (s)-dependent precision B_s. And again, T_s is the target parameter that serves as the "attractor" drawing the time-varying hidden dynamic variable toward it within each phonological unit denoted by s.

It is easy to verify that this second-order state equation, as for the first-order one, has the desirable properties of target directedness and monotonicity. However, the trajectory implied by the second-order recursion is more realistic than that by the earlier first-order one. The new trajectory has critically damped trajectory shaping, while the first-order trajectory has exponential shaping. Detailed behaviors of the respective trajectories are controlled by the parameter r_s in both the cases. For analysis of such behaviors, see [33, 54].

The explicit probabilistic form of the state equation (4.34) is

$$p(x_t \mid x_{t-1}, x_{t-2}, s_t = s) = N(x_t; 2 r_s x_{t-1} - r_s^2 x_{t-2} + (1 - r_s)^2 T_s, B_s). \tag{4.35}$$

Note the conditioning event is both x_{t-1} and x_{t-2}, instead of just x_{t-1} as in the first-order case.

After discretization of the hidden dyanmic variables x_t, x_{t-1}, and x_{t-2}, Eq. (4.35) turns into an approximate form:

$$p(x_t[i] \mid x_{t-1}[j], x_{t-2}[k], s_t = s) \approx N(x_t[i]; 2r_s x_{t-1}[j] - r_s^2 x_{t-2}[k] + (1 - r_s)^2 T_s, B_s).$$

$$(4.36)$$

4.2.2   Extension from Linear to Nonlinear Mapping

The second step of extension of the basic model involves changing from the linear form of the observation equation

$$o_t = H_s x_t + h_s + v_t,$$

to the new nonlinear form

$$o_t = F(x_t) + h_s + v_t(s),$$ (4.37)

where the output of nonlinear predictive or mapping function $F(x_t)$ is the acoustic measurement that can be computed directly from the speech waveform. The expression $h_s + v_t(s)$ is the prediction residual, where h_s is the state-dependent mean and the observation noise $v_k(s) \sim N(v_k; 0, D_s)$ is an IID, zero-mean Gaussian with precision D_s. The phonological unit or state s in h_s may be further subdivided into several left-to-right subunit states. In this case, we can treat all the state labels s as the subphone states but tie the subphone states in the state equation so that the sets of T_s, r_s, B_s are the same for a given phonological unit. This will simplify the exposition of the model in this section without having to distinguish the state from the "substate" and we will use the same label s to denote both. The nonlinear function $F(x_t)$ may be made phonological-unit-dependent to increase the model discriminability (as in [24]). But for simplicity, we assume that in this chapter it is independent of phonological units.

Again, we rewrite Eq. (4.37) in an explicit probabilistic form of

$$p(o_t \mid x_t, s_t = s) = N(o_t; F(x_t) + h_s, D_s).$$ (4.38)

After discretizing the hidden dyanmic variable x_t, the observation equation (4.38) is approximated by

$$p(o_t \mid x_t[i], s_t = s) \approx N(o_t; F(x_t[i]) + h_s, D_s).$$ (4.39)

Combining this with Eq. (4.35), we have the joint probability model:

$$p(s_1^N, x_1^N, o_1^N) = \prod_{t=1}^{N} \pi_{s_{t-1}s_t} p(x_t \mid x_{t-1}, x_{t-2}, s_t) p(o_t \mid x_t, s_t = s)$$

$$\approx \prod_{t=1}^{N} \pi_{s_{t-1}s_t} N(x[i_t]; 2r_s x[i_{t-1}] - r_s^2 x[i_{t-2}] + (1 - r_s)^2 T_s, B_s)$$
$$\times N(o_t; F(x[i_t]) + h_s, D_s),$$ (4.40)

where i_t, i_{t-1}, and i_{t-2} denote the discretization indices of the hidden dynamic variables at time frames t, $t-1$, and $t-2$, respectively.

4.2.3   An Analytical Form of the Nonlinear Mapping Function

The choice of the functional form of $F(x_t)$ in Eq. (4.38) is critical for the success of the model in applications. In Chapter 2, we discussed the use of neural network functions (MLP and RBF, etc.) as well as the use of piecewise linear functions to represent or approximate the generally nonlinear function responsible for mapping from the hidden dynamic variables to acoustic observation variables. These techniques, while useful as shown in [24, 84, 85, 108, 118], either require a large number of parameters to train, or necessitate crude approximation as needed for carrying out parameter estimation algorithm development.

   In this section, we will present a specific form of the nonlinear function of $F(x)$ that contains no free parameters and that after discretizing the input argument x invokes no further approximation in developing and implementing the EM-based parameter estimation algorithm. The key to developing this highly desirable form of the nonlinear function is to endow the hidden dynamic variables with their physical meaning. In this case, we let the hidden dynamic variables be vocal tract resonances (VTRs, and sometimes called formants) including both resonance frequencies and bandwidths. Then, under reasonable assumptions, we can derive an explicit nonlinear functional relationship between the hidden dynamic variables (in the form of VTRs) and the acoustic observation variables in the form of linear cepstra [5]. We now describe this approach in detail.

Definition of Hidden Dynamic Variables and Related Notations

Let us define the hidden dynamic variables for each frame of speech as the $2K$-dimensional vector of VTRs. It consists of a set of P resonant frequencies $\mathbf{f}$ and corresponding bandwidths $\mathbf{b}$, which we denote as

$$\mathbf{x} = \begin{pmatrix} \mathbf{f} \\ \mathbf{b} \end{pmatrix},$$

where

$$\mathbf{f} = \begin{pmatrix} f_1 \\ f_2 \\ \vdots \\ f_P \end{pmatrix} \quad \text{and} \quad \mathbf{b} = \begin{pmatrix} b_1 \\ b_2 \\ \vdots \\ b_P \end{pmatrix}.$$

We desire to establish a memoryless mapping relationship between the VTR vector $\mathbf{x}$ and an acoustic measurement vector $\mathbf{o}$:

$$\mathbf{o} \approx F(\mathbf{x}).$$

Depending on the type of the acoustic measurements as the output in the mapping function, closed-form computation for $F(\mathbf{x})$ may be impossible, or its in-line computation may be too expensive. To overcome these difficulties, we may quantize each dimension of $\mathbf{x}$ over a range of frequencies or bandwidths, and then compute $C(\mathbf{x})$ for every quantized vector value of $\mathbf{x}$. This will be made especially effective when a closed form of the nonlinear function can be established. We will next show that when the output of the nonlinear function becomes linear cepstra, a closed form can be easily derived.

Derivation of a Closed-form Nonlinear Function from VTR to Cepstra

Consider an all-pole model of speech, with each of its poles represented as a frequency–bandwidth pair (f_p, b_p). Then the corresponding complex root is given by [119]

$$z_p = e^{-\pi \frac{b_p}{f_{\text{samp}}} + j2\pi \frac{f_p}{f_{\text{samp}}}}, \quad \text{and} \quad z_p^* = e^{-\pi \frac{b_p}{f_{\text{samp}}} - j2\pi \frac{f_p}{f_{\text{samp}}}}, \tag{4.41}$$

where f_{samp} is the sampling frequency. The transfer function with P poles and a gain of G is

$$H(z) = G \prod_{p=1}^{P} \frac{1}{(1 - z_p z^{-1})(1 - z_p^* z^{-1})}. \tag{4.42}$$

Taking logarithm on both sides of Eq. (4.42), we obtain

$$\log H(z) = \log G - \sum_{p=1}^{P} \log(1 - z_p z^{-1}) - \sum_{p=1}^{P} \log(1 - z_p^* z^{-1}). \tag{4.43}$$

Now using the well-known infinite series expansion formula

$$\log(1 - v) = -\sum_{n=1}^{\infty} \frac{v^n}{n}, \quad |v| \le 1,$$

and with $v = z_p z^{-1}$, we obtain

$$\log H(z) = \log G + \sum_{p=1}^{P} \sum_{n=1}^{\infty} \frac{z_p^n z^{-n}}{n} + \sum_{p=1}^{P} \sum_{n=1}^{\infty} \frac{z_p^{*n} z^{-n}}{n} = \log G + \sum_{n=1}^{\infty} \left[\sum_{p=1}^{P} \frac{z_p^n + z_p^{*n}}{n} \right] z^{-n}. \tag{4.44}$$

Comparing Eq. (4.44) with the definition of the one-sided z-transform,

$$C(z) = \sum_{n=0}^{\infty} c_n z^{-n} = c_0 + \sum_{n=1}^{\infty} c_n z^{-n},$$

we immediately see that the inverse z-transform of $\log H(z)$ in Eq. (4.44), which by definition is the linear cepstrum, is

$$c_n = \sum_{p=1}^{P} \frac{z_p^n + z_p^{*n}}{n}, \quad n > 0, \tag{4.45}$$

and $c_0 = \log G$.

Using Eq. (4.41) to expand and simplify Eq. (4.45), we obtain the final form of the nonlinear function (for $n > 0$):

$$c_n = \frac{1}{n} \sum_{p=1}^{P} \left[e^{-\pi n \frac{b_p}{f_s} + j 2\pi n \frac{f_p}{f_s}} + e^{-\pi n \frac{b_p}{f_s} - j 2\pi n \frac{f_p}{f_s}} \right]$$

$$= \frac{1}{n} \sum_{p=1}^{P} e^{-\pi n \frac{b_p}{f_s}} \left[e^{j 2\pi n \frac{f_p}{f_s}} + e^{-j 2\pi n \frac{f_p}{f_s}} \right]$$

$$= \frac{1}{n} \sum_{p=1}^{P} e^{-\pi n \frac{b_p}{f_s}} \left[\cos\left(2\pi n \frac{f_p}{f_s}\right) + j \sin\left(2\pi n \frac{f_p}{f_s}\right) + \cos\left(2\pi n \frac{f_p}{f_s}\right) - j \sin\left(2\pi n \frac{f_p}{f_s}\right) \right]$$

$$= \frac{2}{n} \sum_{p=1}^{P} e^{-\pi n \frac{b_p}{f_s}} \cos\left(2\pi n \frac{f_p}{f_s}\right). \tag{4.46}$$

Here, c_n constitutes each of the elements in the vector-valued output of the nonlinear function $F(x)$.

Illustrations of the Nonlinear Function

Equation (4.46) gives the decomposition property of the linear cepstrum—it is a sum of the contributions from separate resonances without interacting with each other. The key advantage of the decomposition property is that it makes the optimization procedure highly efficient for inverting the nonlinear function from the acoustic measurement to the VTR. For details, see a recent publication in [110].

As an illustration, in Figs. 4.1–4.3, we plot the value of one term,

$$e^{-\pi n \frac{b}{f_s}} \cos\left(2\pi n \frac{f}{f_s}\right),$$

in Eq. (4.46) as a function of the resonance frequency f and bandwidth b, for the first-order ($n = 1$), second-order ($n = 2$), and the fifth-order ($n = 5$) cepstrum, respectively. (The sampling frequency $f_s = 8000$ Hz is used in all the plots.) These are the cepstra corresponding to the transfer function of a single-resonance (i.e., one pole with no zeros) linear system. Due to

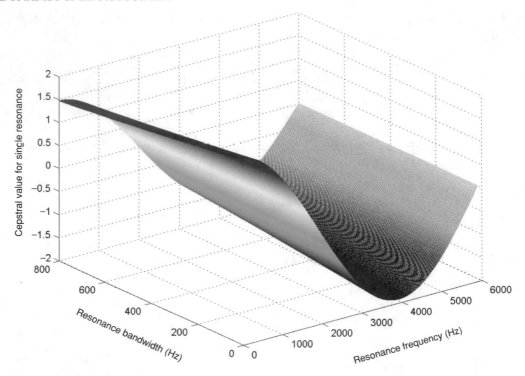

FIGURE 4.1: First-order cepstral value of a one-pole (single-resonance) filter as a function of the resonance frequency and bandwidth. This plots the value of one term in Eq. (4.46) vs. f_p and b_p with fixed $n = 1$ and $f_s = 8000$ Hz

the decomposition property of the linear cepstrum, for multiple-resonance systems, the corresponding cepstrum is simply a sum of those for the single-resonance systems.

Examining Figs. 4.1–4.3, we easily observe some key properties of the (single-resonance) cepstrum. First, the mapping function from the VTR frequency and bandwidth variables to the cepstrum, while nonlinear, is well behaved. That is, the relationship is smooth, and there is no sharp discontinuity. Second, for a fixed resonance bandwidth, the frequency of the sinusoidal relation between the cepstrum and the resonance frequency increases as the cepstral order increases. The implication is that when piecewise linear functions are to be used to approximate the nonlinear function of Eq. (4.46), more "pieces" will be needed for the higher-order than for the lower-order cepstra. Third, for a fixed resonance frequency, the dependence of the low-order cepstral values on the resonance bandwidth is relatively weak. The cause of this weak dependence is the low ratio of the bandwidth (up to 800 Hz) to the sampling frequency (e.g., 16 000 Hz) in the exponent of the cepstral expression in Eq. (4.46). For example, as shown in Fig. 4.1 for the first-order cepstrum, the extreme values of bandwidths from 20 to 800 Hz

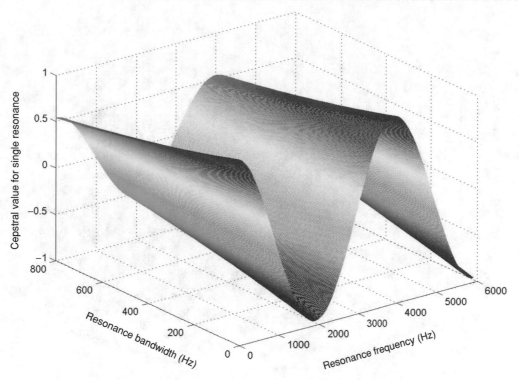

FIGURE 4.2: Second-order cepstral value of a one-pole (single-resonance) filter as a function of the resonance frequency and bandwidth ($n = 1$ and $f_s = 8000$ Hz)

reduce the peak cepstral values only from 1.9844 to 1.4608 (computed by 2 exp($-20\pi/8000$) and 2 exp($-800\pi/8000$), respectively). The corresponding reduction for the second-order cepstrum is from 0.9844 to 0.5335 (computed by exp($-2 \times 20\pi/8000$) and exp($-2 \times 800\pi/8000$), respectively). In general, the exponential decay of the cepstral value, as the resonance bandwidth increases, becomes only slightly more rapid for the higher-order than for the lower-order cepstra (see Fig. 4.3). This weak dependence is desirable since the VTR bandwidths are known to be highly variable with respect to the acoustic environment [120], and to be less correlated with the phonetic content of speech and with human speech perception than are the VTR frequencies.

Quantization Scheme for the Hidden Dynamic Vector

In the discretized hidden dynamic model, which is the theme of this chapter, the discretization scheme is a central issue. We address this issue here using the example of the nonlinear function discussed above, based on the recent work published in [110]. In that work, four poles are used in the LPC model of speech [i.e., using $P = 4$ in Eq. (4.46)], since these lowest VTRs carry the

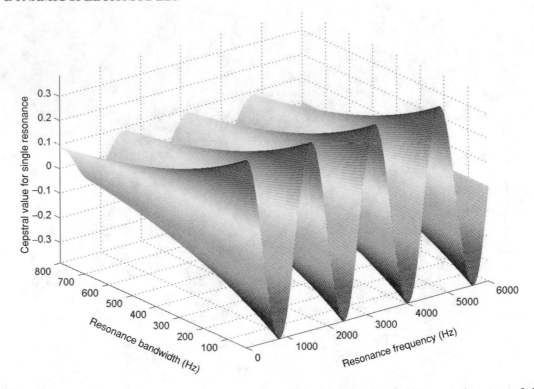

FIGURE 4.3: Fifth-order cepstral value of a one-pole (single-resonance) filter as a function of the resonance frequency and bandwidth $n = 5$ and $f_s = 8000$ Hz

most important phonetic information of the speech signal. That is, an eight-dimensional vector $x = (f_1, f_2, f_3, f_4, b_1, b_2, b_3, b_4)$ is used as the input to the nonlinear function $F(x)$. For the output of the nonlinear function, up to 15 orders of linear cepstra are used. The zeroth order cepstrum, c_0, is excluded from the output vector, making the nonlinear mapping from VTRs to cepstra independent of the energy level in the speech signal. This corresponds to setting the gain $G = 1$ in the all-pole model of Eq. (4.42).

For each of the eight dimensions in the VTR vector, scalar quantization is used. Since $F(x)$ is relevant to all possible phones in speech, the appropriate range is chosen for each VTR frequency and its corresponding bandwidth to cover all phones according to the considerations discussed in [9]. Table 4.1 lists the range, from minimal to maximal frequencies in Hz, for each of the four VTR frequencies and bandwidths. It also lists the corresponding number of quantization levels used. Bandwidths are quantized uniformly with five levels while frequencies are mapped to the Mel-frequency scale and then uniformly quantized with 20 levels. The total number of quantization levels shown in Table 4.1 yields a total of 100 million ($20^4 \times 5^4$)

TABLE 4.1: Quantization Scheme for the VTR Variables, Including the Ranges of the Four VTR Frequencies and Bandwidths and the Corresponding Numbers of Quantization Levels

	MINIMUM (Hz)	MAXIMUM (Hz)	NO. OF QUANTIZATION
f_1	200	900	20
f_2	600	2800	20
f_3	1400	3800	20
f_4	1700	5000	20
b_1	40	300	5
b_2	60	300	5
b_3	60	500	5
b_4	100	700	5

entries for $F(x)$, but because of the constraint $f_1 < f_2 < f_3 < f_4$, the resulting number has been reduced by about 25%.

4.2.4   E-Step for Parameter Estimation

After giving a comprehensive example above for the construction of a vector-valued nonlinear mapping function and the quantization scheme for the vector valued hidden dynamics as the input, we now return to the problem of parameter learning for the extended model. We also return to the scalar case for the purpose of simplicity in exposition. We first describe the E-step in the EM algorithm for the extended model, and concentrate on the differences from the basic model as presented in a greater detail in the preceding section.

Like the basic model, before discretization, the auxiliary function for the E-step can be simplified into the same form of

$$Q(r_s, T_s, B_s, b_s, D_s) = Q_x(r_s, T_s, B_s) + Q_o(b_s, D_s) + \text{Const.}, \tag{4.47}$$

where

$$Q_x(r_s, T_s, B_s) = 0.5 \sum_{s=1}^{S} \sum_{t=1}^{N} \sum_{i=1}^{C} \sum_{j=1}^{C} \sum_{k=1}^{C} \xi_t(s, i, j, k) \Big[\log |B_s| \\ - B_s \big((x_t[i] - 2r_s x_{t-1}[j] + r_s^2 x_{t-2}[k] - (1 - r_s)^2 T_s)^2 \big) \Big], \tag{4.48}$$

and

$$Q_o(b_s, D_s) = 0.5 \sum_{s=1}^{S} \sum_{t=1}^{N} \sum_{i=1}^{C} \gamma_t(s, i) \Big[\log |D_s| - D_s \big(o_t - F(x_t[i]) - b_s \big)^2 \Big]. \tag{4.49}$$

Again, large computational saving can be achieved by limiting the summations in Eq. (4.48) for i, j, k based on the relative smoothness of trajectories in x_t. That is, the range of i, j, k can be set such that $|x_t[i] - x_{t-1}[j]| < \mathrm{Th}_1$, and $|x_{t-1}[j] - x_{t-2}[k]| < \mathrm{Th}_2$. Now two thresholds, instead of one in the basic model, are to be set.

In the above, we used $\xi_t(s, i, j, k)$ and $\gamma_t(s, i)$ to denote the frame-level posteriors of

$$\xi_t(s, i, j, k) \equiv p(s_t = s, x_t[i], x_{t-1}[j], x_{t-2}[k] \mid o_1^N),$$

and

$$\gamma_t(s, i) \equiv p(s_t = s, x_t[i] \mid o_1^N).$$

Note that $\xi_t(s, i, j, k)$ has one more index k than the counterpart in the basic model. This is due to the additional conditioning in the second-order state equation.

Similar to the basic model, in order to compute $\xi_t(s, i, j, k)$ and $\gamma_t(s, i)$, we need to compute the forward and backward probabilities by recursion. The forward recursion $\alpha_t(s, i) \equiv p(o_1^t, s_t = s, i_t = i)$ is

$$\alpha(s_{t+1}, i_{t+1}) = \sum_{s_t=1}^{S} \sum_{i_t=1}^{C} \alpha(s_t, i_t) p(s_{t+1}, i_{t+1} \mid s_t, i_t, i_{t-1}) p(o_{t+1} \mid s_{t+1}, i_{t+1}), \qquad (4.50)$$

where

$$p(o_{t+1} \mid s_{t+1} = s, i_{t+1} = i) = N(o_{t+1}; F(x_{t+1}[i]) + h_s, D_s),$$

and

$$\begin{aligned}
&p(s_{t+1} = s, i_{t+1} = i \mid s_t = s', i_t = j, i_{t-1} = k) \\
&\approx p(s_{t+1} = s \mid s_t = s') p(i_{t+1} = i \mid i_t = j, i_{t-1} = k) \\
&= \pi_{s's} N(x_t[i]; 2r_s x_{t-1}[j] - r_s^2 x_{t-2}[k] + (1 - r_s)^2 T_s, B_s).
\end{aligned}$$

The backward recursion $\beta_t(s, i) \equiv p(o_{t+1}^N \mid s_t = s, i_t = i)$ is

$$\beta(s_t, i_t) = \sum_{s_{t+1}=1}^{S} \sum_{i_{t+1}=1}^{C} \beta(s_{t+1}, i_{t+1}) p(s_{t+1}, i_{t+1} \mid s_t, i_t, i_{t-1}) p(o_{t+1} \mid s_{t+1}, i_{t+1}). \qquad (4.51)$$

Given $\alpha_t(s, i)$ and $\beta(s_t, i_t)$ as computed, we can obtain the posteriors of $\xi_t(s, i, j, k)$ and $\gamma_t(s, i)$.

4.2.5   M-Step for Parameter Estimation

Reestimation for Parameter r_s

To obtain the reestimation formula for parameter r_s, we set the following partial derivative to zero:

$$\frac{\partial Q_x(r_s, T_s, B_s)}{\partial r_s} = -B_s \sum_{t=1}^{N} \sum_{i=1}^{C} \sum_{j=1}^{C} \sum_{k=1}^{C} \xi_t(s, i, j, k) \tag{4.52}$$

$$\times \left[x_t[i] - 2r_s x_{t-1}[j] + r_s^2 x_{t-2}[k] - (1 - r_s)^2 T_s \right] \left[-x_{t-1}[j] + r_s x_{t-2}[k] + (1 - r_s)T_s \right]$$

$$= -B_s \sum_{t=1}^{N} \sum_{i=1}^{C} \sum_{j=1}^{C} \sum_{k=1}^{C} \xi_t(s, i, j, k)$$

$$\times \Big[-x_t[i]x_{t-1}[j] + 2r_s x_{t-1}^2[j] - r_s^2 x_{t-1}[j]x_{t-2}[k] + (1 - r_s)^2 x_{t-1}[j]T_s$$

$$+ r_s x_t[i]x_{t-2}[k] - 2r_s^2 x_{t-1}[j]x_{t-2}[k] + r_s^3 x_{t-2}^2[k] - r_s(1 - r_s)^2 x_{t-2}[k]T_s$$

$$+ x_t[i](1 - r_s)T_s - 2r_s x_{t-1}[j](1 - r_s)T_s + r_s^2 x_{t-2}[k](1 - r_s)T_s - (1 - r_s)^3 T_s^2 \Big] = 0.$$

This can be written in the following form in order to solve for r_s (assuming T_s is fixed from the previous EM iteration):

$$A_3 \hat{r}_s^3 + A_2 \hat{r}_s^2 + A_1 \hat{r}_s + A_0 = 0, \tag{4.53}$$

where

$$A_3 = \sum_{t=1}^{N} \sum_{i=1}^{C} \sum_{j=1}^{C} \sum_{k=1}^{C} \xi_t(s, i, j, k)\{x_{t-2}^2[k] + T_s x_{t-2}[k] + T_s^2\},$$

$$A_2 = \sum_{t=1}^{N} \sum_{i=1}^{C} \sum_{j=1}^{C} \sum_{k=1}^{C} \xi_t(s, i, j, k)\{-3x_{t-1}[j]x_{t-2}[k] + 3T_s x_{t-1}[j] + 3T_s x_{t-2}[k] - 3T_s^2\},$$

$$A_1 = \sum_{t=1}^{N} \sum_{i=1}^{C} \sum_{j=1}^{C} \sum_{k=1}^{C} \xi_t(s, i, j, k)\{2x_{t-1}^2[j] + x_t[i]x_{t-2}[k] - x_t[i]T_s$$

$$- 4x_{t-1}[j]T_s - x_{t-2}[k]T_s + 3T_s^2\},$$

$$A_0 = \sum_{t=1}^{N} \sum_{i=1}^{C} \sum_{j=1}^{C} \sum_{k=1}^{C} \xi_t(s, i, j, k)\{-x_t[i]x_{t-1}[j] + x_t[i]T_s + x_{t-1}[j]T_s - T_s^2\}. \tag{4.54}$$

Analytic solutions exist for third-order algebraic equations such as the above. For the three roots found, constraints $1 > r_s > 0$ can be used for selecting the appropriate one. If there is more than one solution satisfying the constraint, then we can select the one that gives the largest value for Q_x.

Reestimation for Parameter T_s

We now optimize T_s by setting the following partial derivative to zero:

$$\frac{\partial Q_x(r_s, T_s, B_s)}{\partial T_s} = -B_s \sum_{t=1}^{N} \sum_{i=1}^{C} \sum_{j=1}^{C} \sum_{k=1}^{C} \xi_t(s, i, j, k)[x_t[i]$$
$$-2r_s x_{t-1}[j] + r_s^2 x_{t-2}[k] - (1-r_s)^2 T_s](1-r_s)^2 = 0. \quad (4.55)$$

Now fixing r_s from the previous EM iteration, we obtain an explicit solution to the reestimate of T_s:

$$\hat{T}_s = \frac{1}{(1-r_s)^2} \sum_{t=1}^{N} \sum_{i=1}^{C} \sum_{j=1}^{C} \sum_{k=1}^{C} \xi_t(s, i, j, k)\{x_t[i] - 2r_s x_{t-1}[j] + r_s^2 x_{t-2}[k]\}.$$

Reestimation for Parameter h_s

We set

$$\frac{\partial Q_o(h_s, D_s)}{\partial h_s} = -D_s \sum_{t=1}^{N} \sum_{i=1}^{C} \gamma_t(s, i)\{o_t - F(x_t[i]) - h_s\} = 0. \quad (4.56)$$

This gives the reestimation formula:

$$\hat{h}_s = \frac{\sum_{t=1}^{N} \sum_{i=1}^{C} \gamma_t(s, i)\{o_t - F(x_t[i])\}}{\sum_{t=1}^{N} \sum_{i=1}^{C} \gamma_t(s, i)}. \quad (4.57)$$

Reestimation for B_s and D_s

Setting

$$\frac{\partial Q_x(r_s, T_s, B_s)}{\partial B_s} = 0.5 \sum_{t=1}^{N} \sum_{i=1}^{C} \sum_{j=1}^{C} \sum_{k=1}^{C} \xi_t(s, i, j, k)[B_s^{-1}$$
$$- (x_t[i] - 2r_s x_{t-1}[j] + r_s^2 x_{t-2}[k] - (1-r_s)^2 T_s)^2] = 0, \quad (4.58)$$

we obtain the reestimation formula:

$$\hat{B}_s = \frac{\sum_{t=1}^{N} \sum_{i=1}^{C} \sum_{j=1}^{C} \sum_{k=1}^{C} \xi_t(s, i, j, k)\left[x_t[i] - 2r_s x_{t-1}[j] + r_s^2 x_{t-2}[k] - (1-r_s)^2 T_s\right]^2}{\sum_{t=1}^{N} \sum_{i=1}^{C} \sum_{j=1}^{C} \sum_{k=1}^{C} \xi_t(s, i, j, k)}.$$
$$(4.59)$$

Similarly, setting

$$\frac{\partial Q_o(H_s, h_s, D_s)}{\partial D_s} = 0.5 \sum_{t=1}^{N} \sum_{i=1}^{C} \gamma_t(s, i)\left[D_s^{-1} - (o_t - H_s x_t[i] - h_s)^2\right] = 0,$$

we obtain the reestimate (scalar value) of

$$\hat{D}_s = \frac{\sum_{t=1}^{N}\sum_{i=1}^{C}\gamma_t(s,i)[o_t - H_s x_t[i] - h_s]^2}{\sum_{t=1}^{N}\sum_{i=1}^{C}\gamma_t(s,i)}. \tag{4.60}$$

4.2.6   Decoding of Discrete States by Dynamic Programming

The DP recursion is essentially the same as in the basic model, except an additional level (index k) of optimization is introduced due to the second-order dependency in the state equation. The final form of the recursion can be written as

$$\delta_{t+1}(s,i) = \max_{s',i'} \delta_t(s',i')p(s_{t+1}=s, i_{t+1}=i \mid s_t=s', i_t=i')p(o_{t+1}\mid s_{t+1}=s, i_{t+1}=i)$$

$$\approx \max_{s',i'} \delta_t(s',i')p(s_{t+1}=s \mid s_t=s')p(i_{t+1}=i \mid i_t=i')p(o_{t+1}\mid s_{t+1}=s, i_{t+1}=i)$$

$$= \max_{s',i',j,k} \delta_t(s',i')\pi_{s's} N(x_{t+1}[i]; 2r_{s'}x_t[j] - r_{s'}^2 x_{t-1}[k] + (1-r_{s'})^2 T_s, B_s)$$

$$\times N(o_{t+1}; F(x_{t+1}[i]) + h_s, D_s). \tag{4.61}$$

4.3   APPLICATION TO AUTOMATIC TRACKING OF HIDDEN DYNAMICS

As an example for the application of the discretized hidden dynamic model discussed in this chapter so far, we discuss implementation efficiency issues and show results for the specific problem of automatic tracking of the hidden dynamic variables that are discretized. The accuracy of the tracking is obviously limited by the discretization level, but this approximation makes it possible to run the parameter learning and decoding algorithms in a manner that is not only tractable but also efficient.

While the description of the parameter learning and decoding algorithms earlier in this chapter is confined to the scalar case for purposes of clarification and notational convenience, in practical cases where often vector valued hidden dynamics are involved, we need to address the problem of algorithms' efficiency. In the application example in this section where eight-dimensional hidden dynamic vectors (four VTR frequencies and four bandwidths $x = (f_1, f_2, f_3, f_4, b_1, b_2, b_3, b_4)$) are used as presented in detail in Section 4.2.3, it is important to address the issue related to the algorithms' efficiency.

4.3.1   Computation Efficiency: Exploiting Decomposability in the Observation Function

For multidimensional hidden dynamics, one obvious difficulty for the training and tracking algorithms presented earlier is the high computational cost in summing and in searching over the huge space in the quantized hidden dynamic variables. The sum with C terms as required in the various reestimation formulas and in the dynamic programming recursion is typically expensive since C is very large. With scalar quantization for each of the eight VTR dimensions, the C would be the Cartesian product of the quantization levels for each of the dimensions.

To overcome this difficulty, a suboptimal, greedy technique is implemented as described in [110]. This technique capitalizes on the decomposition property of the nonlinear mapping function from VTR to cepstra that we described earlier in Section 4.2.3. This enables a much smaller number of terms to be evaluated than the rigorous number determined as the Cartesian product, which we elaborate below.

Let us consider an objective function F, to be optimized with respect to M noninteracting or decomposable variables that determine the function's value. An example is the following decomposable function consisting of M terms F_m, $m = 1, 2, \ldots, M$, each of which contains independent variables (α_m) to be searched for:

$$F = \sum_{m=1}^{M} F_m(\alpha_m).$$

Note that the VTR-to-cepstrum mapping function, which was derived to be Eq. (4.46) as the observation equation of the dynamic speech model (extended model), has this decomposable form. The greedy optimization technique proceeds as follows. First, initialize α_m, $m = 1, 2, \ldots, M$ to reasonable values. Then, fix all $\alpha'_m s$ except one, say α_n, and optimize α_n with respect to the new objective function of

$$F - \sum_{m=1}^{n-1} F_m(\alpha_m) - \sum_{m=n+1}^{M} F_m(\alpha_m).$$

Next, after the low-dimensional, inexpensive search problem for $\hat{\alpha}_n$ is solved, fix it and optimize a new α_m, $m \neq n$. Repeat this for all $\alpha'_m s$. Finally, iterate the above process until all optimized $\alpha'_m s$ become stabilized.

In the implementation of this technique for VTR tracking and parameter estimation as reported in [110], each of the $P = 4$ resonances is treated as a separate, noninteractive variables to optimize. It was found that only two to three overall iterations above are already sufficient to stabilize the parameter estimates. (During the training of the residual parameters, these inner iterations are embedded in each of the outer EM iterations.) Also, it was found that initialization of all VTR variables to zero gives virtually the same estimates as those by more carefully thought initialization schemes.

With the use of the above greedy, suboptimal technique instead of full optimal search, the computation cost of VTR tracking was reduced by over 4000-fold compared with the brute-force implementation of the algorithms.

4.3.2   Experimental Results

As reported in [110], the above greedy technique was incorporated into the VTR tracking algorithm and into the EM training algorithm for the nonlinear-prediction residual parameters. The state equation was made simpler than the counterpart in the basic or extended model in that all the phonological states s are tied. This is because for the purposes of tracking hidden dynamics there is no need to distinguish the phonological states. The DP recursion in the more general case of Eq. (4.33) can then be simplified by eliminating the optimization on index s, leaving only the indices i and j of the discretization levels in the hidden VTR variables during the DP recursion. We also set the parameter $r_s = 1$ uniformly in all the experiments. This gives the role of the state equation as a "smoothness" constraint.

The effectiveness of the EM parameter estimation, Eqs. (4.57) and (4.60) in particular, discussed for the extended model in this chapter will be demonstrated in the VTR tracking experiments. Due to the tying of the phonological states, the training does not require any data labeling and is fully unsupervised. Fig. 4.4 shows the VTR tracking (f_1, f_2, f_3, f_4) results,

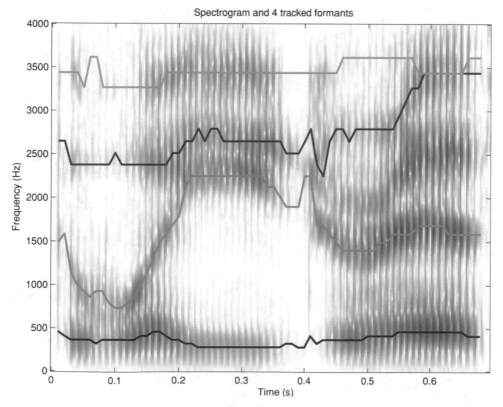

FIGURE 4.4: VTR tracking by setting the residual mean vector to zero

superimposed on the spectrogram of a telephone speech utterance (excised from the Switchboard database) of "*the way you dress*" by a male speaker, when the residual mean vector **h** (tied over all s state) was set to zero and the covariance matrix **D** is set to be diagonal with empirically determined diagonal values. [The initialized variances are those computed from the codebook entries that are constructed from quantizing the nonlinear function in Eq. (4.46.)] Setting **h** to zero corresponds to the assumption that the nonlinear function of Eq. (4.46) is an unbiased predictor of the real speech data in the form of linear cepstra. Under this assumption we observe from Fig. 4.4 that while f_1 and f_2 are accurately tracked through the entire utterance, f_3 and f_4 are incorrectly tracked during the later half of the utterance. (Note that the many small step jumps in the VTR estimates are due to the quantization of the VTR frequencies.) One iteration of the EM training on the residual mean vector and covariance matrix does not correct the errors (see Fig. 4.5), but two iterations are able to correct the errors in the utterance for about 20 frames (after time mark of 0.6 s in Fig. 4.6). One further iteration is able to correct almost all errors as shown in Fig. 4.7.

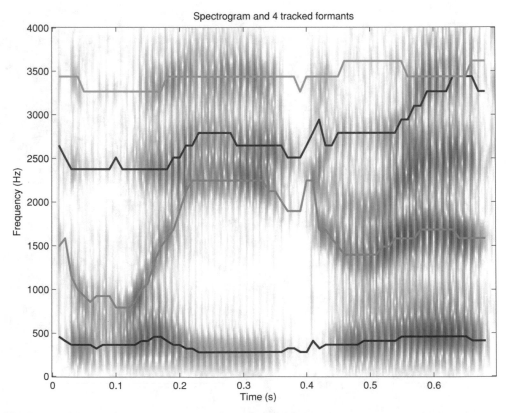

FIGURE 4.5: VTR tracking with one iteration of residual training

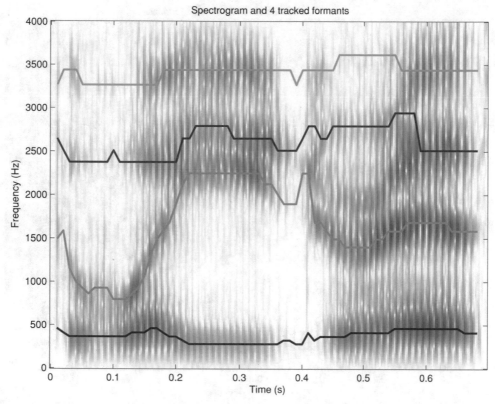

FIGURE 4.6: VTR tracking with two iterations of residual training

To examine the quantitative behavior of the residual parameter training, we list the log-likelihood score as a function of the EM iteration number in Table 4.2. Three iterations of the training appear to have reached the EM convergence. When we examine the VTR tracking results after 5 and 20 iterations, they are found to be identical to Fig. 4.7, consistent with the near-constant converging log-likelihood score reached after three iterations of training. Note that the regions in the utterance where the speech energy is relatively low are where consonantal constriction or closure is formed; e.g., near time mark of 0.1 s for /w/ constriction and near time mark of 0.4 s for /d/ closure). The VTR tracker gives almost as accurate estimates for the resonance frequencies in these regions as for the vowel regions.

4.4   SUMMARY

This chapter discusses one of the two specific types of hidden dynamic models in this book, as example implementations of the general modeling and computational scheme introduced in Chapter 2. The essence of the implementation described in this chapter is the discretization

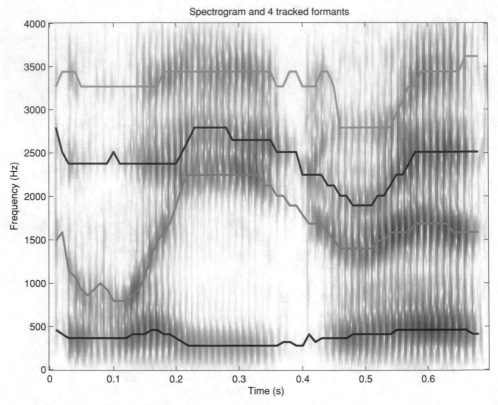

FIGURE 4.7: VTR tracking with three iterations of residual training

TABLE 4.2: Log-likelihood Score as a Function of the EM Iteration Number in Training the Nonlinear-prediction Residual Parameters

EM ITERATION NO.	LOG-LIKELIHOOD SCORE
0	1.7680
1	2.0813
2	2.0918
3	2.1209
5	2.1220
20	2.1222

of the hidden dynamic variables. While this implementation introduces approximations to the original continuous-valued variables, the otherwise intractable parameter estimation and decoding algorithms have become tractable, as we have presented in detail in this chapter.

This chapter starts by introducing the "basic" model, where the state equation in the dynamic speech model gives discretized first-order dynamics and the observation equation is a linear relationship between the discretized hidden dynamic variables and the acoustic observation variables. Probabilistic formulation of the model is presented first, which is equivalent to the state–space formulation but is in a form that can be more readily used in developing and describing the model parameter estimation algorithms. The parameter estimation algorithms are presented, with sufficient detail in deriving all the final reestimation formulas as well as the key intermediate quantities such as the auxiliary function in the E-step of the EM algorithm. In particular, we separate the forward–backward algorithm out of the general E-step derivation in a new subsection to emphasize its critical role. After deriving the reestimation formulas for all model parameters as the M-step, we describe a DP-based algorithm for jointly decoding the discrete phonological states and the hidden dynamic "state," the latter constructed from discretization of the continuous variables.

The chapter is followed by presenting an extension of the basic model in two aspects. First, the state equation is extended from the first-order dynamics to the second-order dynamics, making the shape of the temporally unfolded "trajectories" more realistic. Second, the observation equation is extended from the linear mapping to a nonlinear one. A new subsection is then devoted to a special construction of the nonlinear mapping where a "physically" based prediction function is developed when the hidden dynamic variables as the input are taken to be the VTRs and the acoustic observations as the output are taken to be the linear cepstral features. Using this nonlinear mapping function, we proceed to develop the E-step and M-step of the EM algorithm for this extended model in a way parallel to that for the basic model.

Finally, we give an application example of the use of a simplified version of the extended model and the related algorithms discussed in this chapter for automatic tracking of the hidden dynamic vectors, the VTR trajectories in this case. Specific issues related to the tracking algorithm's efficiency arising from multidimensionality in the hidden dynamics are addressed, and experimental results on some typical outputs of the algorithms are presented and analyzed.

CHAPTER 5

Models with Continuous-Valued Hidden Speech Trajectories

The preceding chapter discussed the implementation strategy for hidden dynamic models based on discretizing the hidden dynamic values. This permits tractable but approximate learning of the model parameters and decoding of the discrete hidden states (both phonological states and discretized hidden dynamic "states"). This chapter elaborates on another implementation strategy where the continuous-valued hidden dynamics remain unchanged but a different type of approximation is used. This implementation strategy assumes fixed discrete-state (phonological unit) boundaries, which may be obtained initially from a simpler speech model set such as the HMMs and then be further refined after the dynamic model is learned iteratively. We will describe this new implementation and approximation strategy for a hidden trajectory model (HTM) where the hidden dynamics are defined as an explicit function of time instead of by recursion. Other types of approximation developed for the recursively defined dynamics can be found in [84, 85, 121–123] and will not be described in this book.

This chapter extracts, reorganizes, and expands the materials published in [109, 115, 116, 124], fitting these materials into the general theme of dynamic speech modeling in this book.

5.1 OVERVIEW OF THE HIDDEN TRAJECTORY MODEL

As a special type of the hidden dynamic model, the HTM presented in this section is a structured generative model, from the top level of phonetic specification to the bottom level of acoustic observations via the intermediate level of (nonrecursive) FIR-based target filtering that generates hidden VTR trajectories. One advantage of the FIR filtering is its natural handling of the two constraints (segment-bound monotonicity and target-directedness) that often requires asynchronous segment boundaries for the VTR dynamics and for the acoustic observations.

This section is devoted to the mathematical formulation of the HTM as a statistical generative model. Parameterization of the model is detailed here, with consistent notations set up to facilitate the derivation and description of algorithmic learning of the model parameters presented in the next section.

5.1.1   Generating Stochastic Hidden Vocal Tract Resonance Trajectories

The HTM assumes that each phonetic unit is associated with a multivariate distribution of the VTR targets. (There are exceptions for several compound phonetic units, including diphthongs and affricates, where two distributions are used.) Each phone-dependent target vector, t_s, consists of four low-order resonance frequencies appended by their corresponding bandwidths, where s denotes the segmental phone unit. The target vector is a random vector—hence stochastic target—whose distribution is assumed to be a (gender-dependent) Gaussian:

$$p(t \mid s) = \mathcal{N}(t; \mu_{T_s}, \Sigma_{T_s}). \tag{5.1}$$

The generative process in the HTM starts by temporal filtering the stochastic targets. This results in a time-varying pattern of stochastic hidden VTR vectors $z(k)$. The filter is constrained so that the smooth temporal function of $z(k)$ moves segment-by-segment towards the respective target vector t_s but it may or may not reach the target depending on the degree of phonetic reduction.

These phonetic targets are segmental in that they do not change over the phone segment once the sample is taken, and they are assumed to be largely context-independent. In our HTM implementation, the generation of the VTR trajectories from the segmental targets is through a bidirectional finite impulse response (FIR) filtering. The impulse response of this noncausal filter is

$$h_s(k) = \begin{cases} c\gamma_{s(k)}^{-k} & -D < k < 0, \\ c & k = 0, \\ c\gamma_{s(k)}^{k} & 0 < k < D, \end{cases} \tag{5.2}$$

where k represents time frame (typically with a length of 10 ms each), and $\gamma_{s(k)}$ is the segment-dependent "stiffness" parameter vector, one component for each resonance. Each component is positive and real-valued, ranging between zero and one. In Eq. (5.2), c is a normalization constant, ensuring that $h_s(k)$ sums to one over all time frames k. The subscript $s(k)$ in $\gamma_{s(k)}$ indicates that the stiffness parameter is dependent on the segment state $s(k)$, which varies over time. D in Eq. (5.2) is the unidirectional length of the impulse response, representing the temporal extent of coarticulation in one temporal direction, assumed for simplicity to be equal in length for the forward direction (anticipatory coarticulation) and the backward direction (regressive coarticulation).

In Eq. (5.2), c is the normalization constant to ensure that the filter weights add up to one. This is essential for the model to produce target undershooting, instead of overshooting. To determine c, we require that the filter coefficients sum to one:

$$\sum_{k=-D}^{D} h_s(k) = c \sum_{k=-D}^{D} \gamma_{s(k)}^{|k|} = 1. \tag{5.3}$$

For simplicity, we make the assumption that over the temporal span of $-D \leq k \leq D$, the stiffness parameter's value stays approximately constant

$$\gamma_{s(k)} \approx \gamma.$$

That is, the adjacent segments within the temporal span of $2D+1$ in length that contribute to the coarticulated home segment have the same stiffness parameter value as that of the home segment. Under this assumption, we simplify Eq. (5.3) to

$$c \sum_{k=-D}^{D} \gamma_{s(k)}^{|k|} \approx c[1 + 2(\gamma + \gamma^2 + \cdots + \gamma^D)] = c \frac{1 + \gamma - 2\gamma^{D+1}}{1 - \gamma}.$$

Thus,

$$c(\gamma) \approx \frac{1 - \gamma}{1 + \gamma - 2\gamma^{D+1}}. \qquad (5.4)$$

The input to the above FIR filter as a linear system is the target sequence, which is a function of discrete time and is subject to abrupt jumps at the phone segments' boundaries. Mathematically, the input is represented as a sequence of stepwise constant functions with variable durations and heights:

$$t(k) = \sum_{i=1}^{I} [u(k - k_{s_i}^l) - u(k - k_{s_i}^r)]t_{s_i}, \qquad (5.5)$$

where $u(k)$ is the unit step function, $k_s^r, s = s_1, s_2, \ldots, s_I$ are the right boundary sequence of the segments (I in total) in the utterance, and $k_s^l, s = s_1, s_2, \ldots, s_I$ are the left boundary sequence. Note the constraint on these starting and end times: $k_{s+1}^l = k_s^r$. The difference of the two boundary sequences gives the duration sequence. $t_s, s = s_1, s_2, \ldots, s_I$ are the random target vectors for segment s.

Given the filter's impulse response and the input to the filter as the segmental VTR target sequence $t(k)$, the filter's output as the model's prediction for the VTR trajectories is the convolution between these two signals. The result of the convolution within the boundaries of home segment s is

$$z(k) = h_{s(k)} * t(k) = \sum_{\tau=k-D}^{k+D} c\gamma_{s(\tau)}^{|k-\tau|} t_{s(\tau)}, \qquad (5.6)$$

where the input target vector's value and the filter's stiffness vector's value typically take not only those associated with the current home segment, but also those associated with the adjacent segments. The latter case happens when the time τ in Eq. (5.6) goes beyond the home segment's boundaries, i.e., when the segment $s(\tau)$ occupied at time τ switches from the home segment to an adjacent one.

The linearity between z and t as in Eq. (5.6) and Gaussianity of the target t makes the VTR vector $z(k)$ (at each frame k) a Gaussian as well. We now discuss the parameterization of this Gaussian trajectory:

$$p(z(k) \mid s) = \mathcal{N}[z(k); \mu_{z(k)}, \Sigma_{z(k)}]. \tag{5.7}$$

The mean vector above is determined by the filtering function:

$$\mu_{z(k)} = \sum_{\tau=k-D}^{k+D} c_\gamma \gamma_{s(\tau)}^{|k-\tau|} \mu_{T_{s(\tau)}} = a_k \cdot \mu_T. \tag{5.8}$$

Each fth component of vector $\mu_{z(k)}$ is

$$\mu_{z(k)}(f) = \sum_{l=1}^{L} a_k(l)\mu_T(l, f), \tag{5.9}$$

where L is the total number of phone-like HTM units as indexed by l, and $f = 1, \ldots, 8$ denotes four VTR frequencies and four corresponding bandwidths.

The covariance matrix in Eq. (5.7) can be similarly derived to be

$$\Sigma_{z(k)} = \sum_{\tau=k-D}^{k+D} c_\gamma^2 \gamma_{s(\tau)}^{2|k-\tau|} \Sigma_{T_{s(\tau)}}.$$

Approximating the covariance matrix by a diagonal one for each phone unit l, we represent its diagonal elements as a vector:

$$\sigma_{z(k)}^2 = v_k \cdot \sigma_T^2. \tag{5.10}$$

and the target covariance matrix is also approximated as diagonal:

$$\Sigma_T(l) \approx \begin{bmatrix} \sigma_T^2(l, 1) & 0 & \cdots & 0 \\ 0 & \sigma_T^2(l, 2) & \cdots & 0 \\ \vdots & \vdots & \ddots & \vdots \\ 0 & 0 & \cdots & \sigma_T^2(l, 8) \end{bmatrix}.$$

The fth element of the vector in Eq. (5.10) is

$$\sigma_{z(k)}^2(f) = \sum_{l=1}^{L} v_k(l)\sigma_T^2(l, f). \tag{5.11}$$

In Eqs. (5.8) and (5.10), a_k and v_k are frame (k)-dependent vectors. They are constructed for any given phone sequence and phone boundaries within the coarticulation range ($2D + 1$ frames) centered at frame k. Any phone unit beyond the $2D + 1$ window contributes a zero value to these "coarticulation" vectors' elements. Both $\boldsymbol{a}_k$ and $\boldsymbol{v}_k$ are a function of the phones' identities and temporal orders in the utterance, and are independent of the VTR dimension f.

5.1.2   Generating Acoustic Observation Data

The next generative process in the HTM provides a forward probabilistic mapping or prediction from the stochastic VTR trajectory $\boldsymbol{z}(k)$ to the stochastic observation trajectory $\boldsymbol{o}(k)$. The observation takes the form of linear cepstra. An analytical form of the nonlinear prediction function $\mathcal{F}[\boldsymbol{z}(k)]$ presented here is in the same form as described (and derived) in Section 4.2.3 of Chapter 4 and is summarized here:

$$\mathcal{F}_q(k) = \frac{2}{q} \sum_{p=1}^{P} e^{-\pi q \frac{b_p(k)}{f_{\text{samp}}}} \cos(2\pi q \frac{f_p(k)}{f_{\text{samp}}}), \tag{5.12}$$

where f_{samp} is the sampling frequency, P is the highest VTR order ($P = 4$), and q is the cepstral order.

We now introduce the cepstral prediction's *residual* vector:

$$\mathbf{r}_s(k) = \boldsymbol{o}(k) - \mathcal{F}[\boldsymbol{z}(k)].$$

We model this residual vector as a Gaussian parameterized by residual mean vector $\boldsymbol{\mu}_{r_{s(k)}}$ and covariance matrix $\Sigma_{r_{s(k)}}$:

$$p(\mathbf{r}_s(k) \mid \boldsymbol{z}(k), s) = \mathcal{N}\left[\mathbf{r}_s(k); \boldsymbol{\mu}_{r_{s(k)}}, \Sigma_{r_{s(k)}}\right]. \tag{5.13}$$

Then the conditional distribution of the observation becomes:

$$p(\boldsymbol{o}(k) \mid \boldsymbol{z}(k), s) = \mathcal{N}\left[\boldsymbol{o}(k); \mathcal{F}[\boldsymbol{z}(k)] + \boldsymbol{\mu}_{r_{s(k)}}, \Sigma_{r_{s(k)}}\right]. \tag{5.14}$$

An alternative form of the distribution in Eq. (5.14) is the following "observation equation":

$$\boldsymbol{o}(k) = \mathcal{F}[\boldsymbol{z}(k)] + \boldsymbol{\mu}_{r_{s(k)}} + \boldsymbol{v}_s(k),$$

where the observation noise $\boldsymbol{v}_s(k) \sim \mathcal{N}(\boldsymbol{v}_s; \boldsymbol{0}, \Sigma_{r_{s(k)}})$.

5.1.3   Linearizing Cepstral Prediction Function

To facilitate computing the acoustic observation (linear cepstra) likelihood, it is important to characterize the linear cepstra uncertainty in terms of its conditional distribution on the VTR, and to simplify the distribution to a computationally tractable form. That is, we need to specify and approximate $p(\boldsymbol{o} \mid \boldsymbol{z}, s)$. We take the simplest approach to linearize the nonlinear mean function of $\mathcal{F}[z(k)]$ in Eq. (5.14) by using the first-order Taylor series approximation:

$$\mathcal{F}[z(k)] \approx \mathcal{F}[z_0(k)] + \mathcal{F}'[z_0(k)](z(k) - z_0(k)), \qquad (5.15)$$

where the components of Jacobian matrix $\mathcal{F}'[\cdot]$ can be computed in a closed form of

$$\mathcal{F}'_q[f_p(k)] = -\frac{4\pi}{f_{\text{samp}}} e^{-\pi q \frac{b_p(k)}{f_{\text{samp}}}} \sin\left(2\pi q \frac{f_p(k)}{f_{\text{samp}}}\right), \qquad (5.16)$$

for the VTR frequency components of z, and

$$\mathcal{F}'_q[b_p(k)] = -\frac{2\pi}{f_{\text{samp}}} e^{-\pi q \frac{b_p(k)}{f_{\text{samp}}}} \cos\left(2\pi q \frac{f_p(k)}{f_{\text{samp}}}\right), \qquad (5.17)$$

for the VTR bandwidth components of z. In the current implementation, the Taylor series expansion point $z_0(k)$ in Eq. (5.15) is taken as the tracked VTR values based on the HTM.

Substituting Eq. (5.15) into Eq. (5.14), we obtain the approximate conditional acoustic observation probability where the mean vector μ_{o_s} is expressed as a linear function of the VTR vector z:

$$p(o(k) \mid z(k), s) \approx \mathcal{N}(o(k); \mu_{o_s(k)}, \Sigma_{r_{s(k)}}), \qquad (5.18)$$

where

$$\mu_{o_{s(k)}} = \mathcal{F}'[z_0(k)]z(k) + \left[\mathcal{F}[z_0(k)] - \mathcal{F}'[z_0(k)]z_0(k) + \mu_{r_{s(k)}}\right]. \qquad (5.19)$$

This then permits a closed-form solution for acoustic likelihood computation, which we derive now.

5.1.4 Computing Acoustic Likelihood

An essential aspect of the HTM is its ability to provide the likelihood value for any sequence of acoustic observation vectors $o(k)$ in the form of cepstral parameters. The efficiently computed likelihood provides a natural scoring mechanism comparing different linguistic hypotheses as needed in speech recognition. No VTR values $z(k)$ are needed in this computation as they are treated as the hidden variables. They are marginalized (i.e., integrated over) in the linear cepstra likelihood computation. Given the model construction and the approximation described in the preceding section, the HTM likelihood computation by marginalization can be carried out in a closed form. Some detailed steps of derivation give

$$p(o(k) \mid s) = \int p[o(k) \mid z(k), s] p[z(k) \mid s] \, dz$$

$$\approx \int \mathcal{N}[o(k); \mu_{o_{s(k)}}, \Sigma_{r_{s(k)}}] \, \mathcal{N}[z(k); \mu_{z(k)}, \Sigma_{z(k)}] \, dz$$

$$= \mathcal{N}\left\{ o(k); \bar{\mu}_{o_s}(k), \bar{\Sigma}_{o_s}(k) \right\}, \tag{5.20}$$

where the time (k)-varying mean vector is

$$\bar{\mu}_{o_s}(k) = \mathcal{F}[z_0(k)] + \mathcal{F}'[z_0(k)][a_k \cdot \mu_T - z_0(k)] + \mu_{r_{s(k)}}, \tag{5.21}$$

and the time-varying covariance matrix is

$$\bar{\Sigma}_{o_s}(k) = \Sigma_{r_{s(k)}} + \mathcal{F}'[z_0(k)] \Sigma_z(k) (\mathcal{F}'[z_0(k)])^{\text{Tr}}. \tag{5.22}$$

The final result of Eqs. (5.20)–(5.22) are quite intuitive. For instance, when the Taylor series expansion point is set at $z_0(k) = \mu_z(k) = a_k \cdot \mu_T$, Eq. (5.21) is simplified to $\bar{\mu}_{o_s}(k) = \mathcal{F}[\mu_z(k)] + \mu_{r_s}$, which is the noise-free part of cepstral prediction. Also, the covariance matrix in Eq. (5.20) is increased by the quantity $\mathcal{F}'[z_0(k)] \Sigma_z(k) (\mathcal{F}'[z_0(k)])^{\text{Tr}}$ over the covariance matrix for the cepstral residual term $\Sigma_{r_{s(k)}}$ only. This magnitude of increase reflects the newly introduced uncertainty in the hidden variable, measured by $\Sigma_z(k)$. The variance amplification factor $\mathcal{F}'[z_0(k)]$ results from the local "slope" in the nonlinear function $\mathcal{F}[z]$ that maps from the VTR vector $z(k)$ to cepstral vector $o(k)$.

It is also interesting to interpret the likelihood score Eq. (5.20) as probabilistic characterization of a temporally varying Gaussian process, where the time-varying mean vectors are expressed in Eq. (5.21) and the time-varying covariance matrices are expressed in Eq. (5.22). This may make the HTM look ostensibly like a nonstationary-state HMM (within the acoustic dynamic model category). However, the key difference is that in HTM the dynamic structure represented by the hidden VTR trajectory enters into the time-varying mean vector Eq. (5.21) in two ways: (1) as the argument $z_0(k)$ in the nonlinear function $\mathcal{F}[z_0(k)]$; and (2) as the term $a_k \cdot \mu_T = \mu_{z(k)}$ in Eq. (5.21). Being closely related to the VTR tracks, they both capture long-span contextual dependency, yet with mere context-independent VTR target parameters. Similar properties apply to the time-varying covariance matrices in Eq. (5.22). In contrast, the time-varying *acoustic* dynamic models do not have these desirable properties. For example, the polynomial trajectory model [55, 56, 86] does regression fitting directly on the cepstral data, exploiting no underlying speech structure and hence requiring context dependent polynomial coefficients for representing coarticulation. Likewise, the more recent trajectory model [26] also relies on a very large number of free model parameters to capture acoustic feature variations.

5.2   UNDERSTANDING MODEL BEHAVIOR BY COMPUTER SIMULATION

In this section, we present the model simulation results, extracted from the work published in [109], demonstrating major dynamic properties of the HTM. We further compare these results with the corresponding results from direct measurements of reduction in the acoustic–phonetic literature.

To illustrate VTR frequency or formant target undershooting, we first show the spectrogram of three renditions of a three-segment /iy aa iy/ (uttered by the author of this book) in Fig. 5.1. From left to right, the speaking rate increases and speaking effort decreases, with the durations of the /aa/'s decreasing from approximately 230 to 130 ms. Formant target undershooting for f_1 and f_2 is clearly visible in the spectrogram, where automatically tracked formants are superimposed (as the solid lines) in Fig. 5.1 to aid identification of the formant trajectories. (The dashed lines are the initial estimates, which are then refined to give the solid lines.)

5.2.1   Effects of Stiffness Parameter on Reduction

The same kind of target undershooting for f_1 and f_2 as in Fig. 5.1 is exhibited in the model prediction, shown in Fig. 5.2, where we also illustrate the effects of the FIR filter's stiffness parameter on the magnitude of formant undershooting or reduction. The model prediction is the FIR filter's output for f_1 and f_2. Figs. 5.2(a)–(c) correspond to the use of the stiffness parameter value (the same for each formant vector component) set at $\gamma = 0.85, 0.75$ and 0.65, respectively, where in each plot the slower /iy aa iy/ sounds (with the duration of /aa/ set at

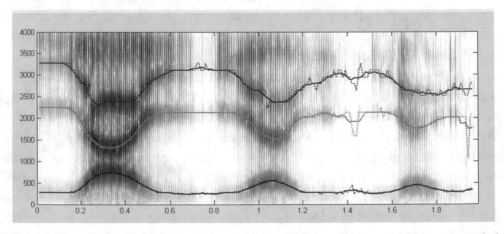

FIGURE 5.1: Spectrogram of three renditions of /iy aa iy/ by one author, with an increasingly higher speaking rate and increasingly lower speaking efforts. The horizontal label is time, and the vertical one is frequency

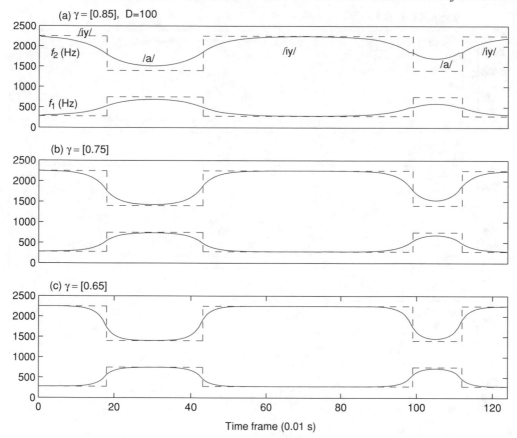

FIGURE 5.2: f_1 and f_2 formant or VTR frequency trajectories produced from the model for a slow /iy aa iy/ followed by a fast /iy aa iy/. (a), (b) and (c) correspond to the use of the stiffness parameter values of $\gamma = 0.85, 0.75$ and 0.65, respectively. The amount of formant undershooting or reduction during the fast /aa/ is decreasing as the γ value decreases. The *dashed lines* indicate the formant target values and their switch at the segment boundaries

230 ms or 23 frames) are followed by the faster /iy aa iy/ sounds (with the duration of /aa/ set at 130 ms or 13 frames). f_1 and f_2 targets for /iy/ and /aa/ are set appropriately in the model also. Comparing the three plots, we have the model's quantitative prediction for the magnitude of reduction in the faster /aa/ that is decreasing as the γ value decreases.

In Figs. 5.3(a)–(c), we show the same model prediction as in Fig. 5.2 but for different sounds /iy eh iy/, where the targets for /eh/ are much closer to those of the adjacent sound /iy/ than in the previous case for /aa/. As such, the absolute amount of reduction becomes smaller. However, the same effect of the filter parameter's value on the size of reduction is shown as for the previous sounds /iy aa iy/.

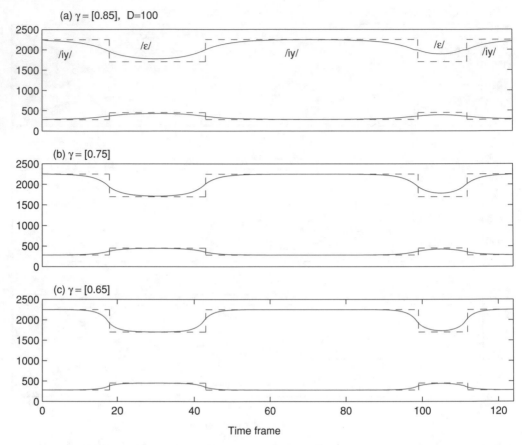

FIGURE 5.3: Same as Fig. 5.2 except for the /iy eh iy/ sounds. Note that the f_1 and f_2 target values for /eh/ are closer to /iy/ than those for /aa/

5.2.2   Effects of Speaking Rate on Reduction

In Fig. 5.4, we show the effects of speaking rate, measured as the inverse of the sound segment's duration, on the magnitude of formant undershooting. Subplots (a)–(c) correspond to three decreasing durations of the sound /aa/ in the /iy aa iy/ sound sequence. They illustrate an increasing amount of the reduction with the decreasing duration or increasing speaking rate. Symbol "x" in Fig. 5.4 indicates the f_1 and f_2 formant values at the central portions of vowels/ aa/, which are predicted from the model and are used to quantify the magnitude of reduction. These values (separately for f_1 and f_2) for /aa/ are plotted against the inversed duration in Fig. 5.5, together with the corresponding values for /eh/ (i.e., IPA ϵ) in the /iy eh iy/ sound sequence. The most interesting observation is that as the speaking rate increases, the distinction between vowels /aa/ and /eh/ gradually diminishes if their static formant values extracted from the dynamic patterns are used as the sole measure for the difference between the sounds. We

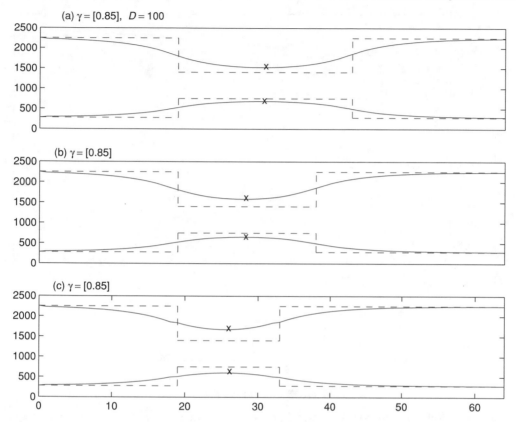

FIGURE 5.4: f_1 and f_2 formant trajectories produced from the model for three different durations of /aa/ in the /iy aa iy/ sounds: (a) 25 frames (250 ms), (b) 20 frames and (c) 15 frames. The same γ value of 0.85 is used. The amount of target undershooting increases as the duration is shortened or the speaking rate is increased. Symbol "x" indicates the f_1 and f_2 formant values at the central portions of vowels of /aa/

refer to this phenomenon as "static" sound confusion induced by increased speaking rate (or/and by a greater degree of sloppiness in speaking).

5.2.3   Comparisons with Formant Measurement Data

The "static" sound confusion between /aa/ and /eh/ quantitatively predicted by the model as shown in Fig. 5.5 is consistent with the formant measurement data published in [125], where thousands of natural sound tokens were used to investigate the relationship between the degree of formant undershooting and speaking rate. We reorganized and replotted the raw data from [125] in Fig. 5.6, in the same formant as Fig. 5.5. While the measures of speaking rate differ between the measurement data and model prediction and cannot be easily converted to each other, they are generally consistent with each other. The similar trend for the greater

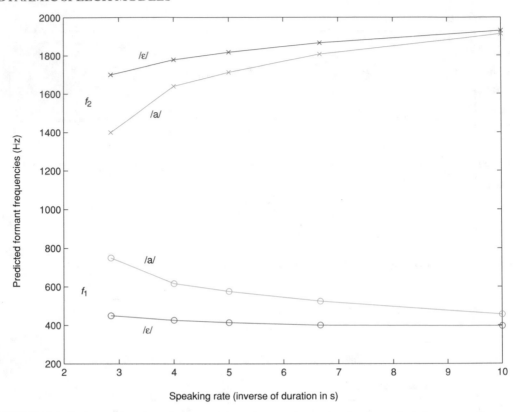

FIGURE 5.5: Relationship, based on model prediction, between the f_1 and f_2 formant values at the central portions of vowels and the speaking rate. Vowel /aa/ is in the carry-phrase /iy aa iy/, and vowel /eh/ in /iy eh iy/. Note that as the speaking rate increases, the distinction between vowels /aa/ and /eh/ measured by the difference between their static formant values gradually diminishes. The same γ value of 0.9 is used in generating all points in the figure

degree of "static" sound confusion as speaking rate increases is clearly evident from both the measurement data (Fig. 5.6) and prediction (Fig. 5.5).

5.2.4   Model Prediction of Vocal Tract Resonance Trajectories for Real Speech Utterances

We have used the expected VTR trajectories computed from the HTM to predict actual VTR frequency trajectories for real speech utterances from the TIMIT database. Only the phone identities and their boundaries are input to the model for the prediction, and no use is made of speech acoustics. Given the phone sequence in any utterance, we first break up the compound phones (affricates and diphthongs) into their constituents. Then we obtain the initial VTR

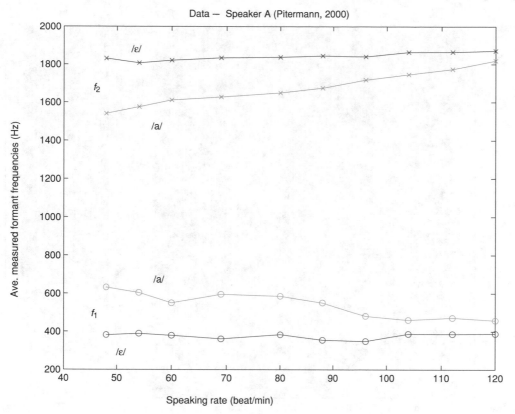

FIGURE 5.6: The formant measurement data from literature are reorganized and plotted, showing similar trends to the model prediction under similar conditions

target values based on limited context dependency by table lookup (see details in [9], Ch. 13). Then automatic and iterative target adaptation is performed for each phone-like unit based on the difference between the results of a VTR tracker (described in [126]) and the VTR prediction from the FIR filter model. These target values are provided not only to vowels, but also to consonants for which the resonance frequency targets are used with weak or no acoustic manifestation. The converged target values, together with the phone boundaries provided from the TIMIT database, form the input to the FIR filter of the HTM and the output of the filter gives the predicted VTR frequency trajectories.

Three example utterances from TIMIT (SI1039, SI1669 and SI2299) are shown in Figs. 5.7–5.9. The stepwise dashed lines ($f_1/f_2/f_3/f_4$) are the target sequences as inputs to the FIR filter, and the continuous lines ($f_1/f_2/f_3/f_4$) are the outputs of the filter as the predicted VTR frequency trajectories. Parameters γ and D are fixed and not automatically learned. To facilitate assessment of the accuracy in the prediction, the inputs and outputs are superimposed

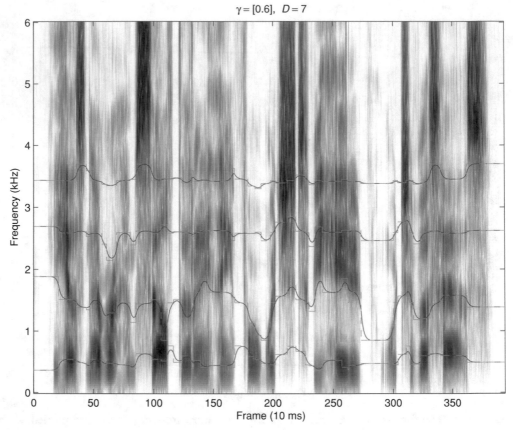

FIGURE 5.7: The $f_1/f_2/f_3/f_4$ VTR frequency trajectories (*smooth lines*) generated from the FIR model for VTR target filtering using the phone sequence and duration of a speech utterance (SI1039) taken from the TIMIT database. The target sequence is shown as *stepwise lines*, switching at the phone boundaries labeled in the database. They are superimposed on the utterance's spectrogram. The utterance is "*He has never, himself, done anything for which to be hated—which of us has*"

on the spectrograms of these utterances, where the true resonances are shown as the dark bands. For the majority of frames, the filter's output either coincides or is close to the true VTR frequencies, even though no acoustic information is used. Also, comparing the input and output of the filter, we observe only a rather mild degree of target undershooting or reduction in these and many other TIMIT utterances we have examined but not shown here.

5.2.5   Simulation Results on Model Prediction for Cepstral Trajectories
The predicted VTR trajectories in Figs. 5.7–5.9 are fed into the nonlinear mapping function in the HTM to produce the predicted linear cepstra shown in Figs. 5.10–5.12, respectively, for

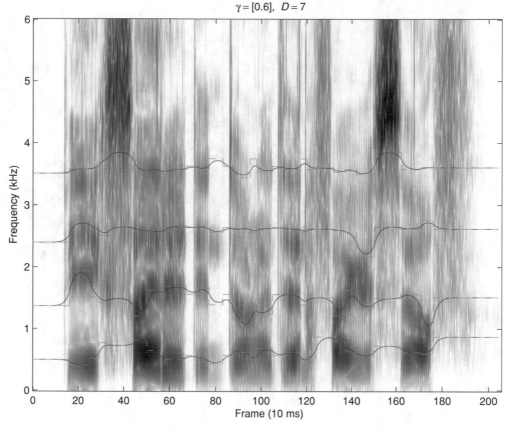

FIGURE 5.8: Same as Fig. 5.7 except with another utterance *"Be excited and don't identify yourself"* (SI1669)

the three example TIMIT utterances. Note that the model prediction includes residual means, which are trained from the full TIMIT data set using an HTK tool. The zero-mean random component of the residual is ignored in these figures. The residual means for the substates (three for each phone) are added sequentially to the output of the nonlinear function Eq. (5.12), assuming each substate occupies three equal-length subsegments of the entire phone segment length provided by TIMIT database. To avoid display cluttering, only linear cepstra with orders one (C1), two (C2) and three (C3) are shown here, as the solid lines. Dashed lines are the linear cepstral data C1, C2 and C3 computed directly from the waveforms of the same utterances for comparison purposes. The data and the model prediction generally agree with each other, somewhat better for lowerorder cepstra than for higherorder ones. It was found that these discrepancies are generally within the variances of the prediction residuals automatically trained from the entire TIMIT training set (using an HTK tool for monophone HMM training).

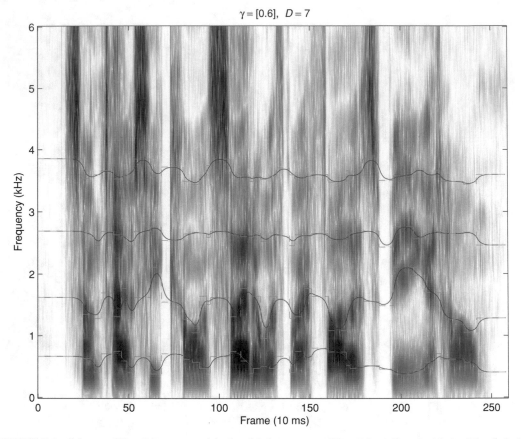

$\gamma = [0.6], \quad D = 7$

FIGURE 5.9: Same as Fig. 5.7 except with the third utterance "*Sometimes, he coincided with my father's being at home*" (SI2299)

5.3   PARAMETER ESTIMATION

In this section, we will present in detail a novel parameter estimation algorithm we have developed and implemented for the HTM described in the preceding section, using the linear cepstra as the acoustic observation data in the training set. The criterion used for this training is to maximize the acoustic observation likelihood in Eq. (5.20). The full set of the HTM parameters consists of those characterizing the linear cepstra residual distributions and those characterizing the VTR target distributions. We present their estimation separately below, assuming that all phone boundaries are given (as in the TIMIT training data set).

5.3.1   Cepstral Residuals' Distributional Parameters

This subset of the HTM parameters consists of (1) the mean vectors $\boldsymbol{\mu}_{r_s}$ and (2) the diagonal elements $\sigma^2_{r_s}$ in the covariance matrices of the cepstral prediction residuals. Both of them are conditioned on phone or sub-phone segmental unit s.

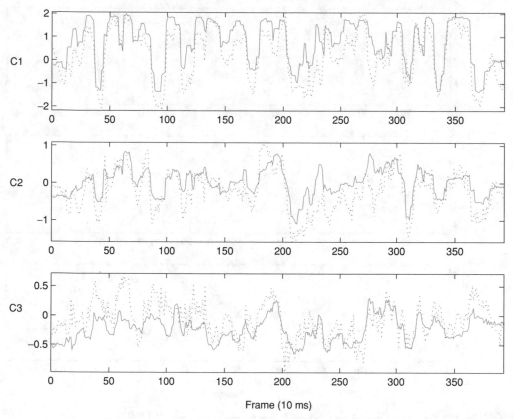

FIGURE 5.10: Linear cepstra with order one (C1), two (C2) and three (C3) predicted from the final stage of the model generating the linear cepstra (*solid lines*) with the input from the FIR filtered results (for utterance SI1039). *Dashed lines* are the linear cepstral data C1, C2 and C3 computed directly from the waveform

Mean Vectors

To find the ML (maximum likelihood) estimate of parameters $\boldsymbol{\mu}_{r_s}$, we set

$$\frac{\partial \log \prod_{k=1}^{K} p(\boldsymbol{o}(k) \mid s)}{\partial \boldsymbol{\mu}_{r_s}} = 0,$$

where $p(\boldsymbol{o}(k) \mid s)$ is given by Eq. (5.20), and K denotes the total duration of sub-phone s in the training data. This gives

$$\sum_{k=1}^{K} \left[\boldsymbol{o}(k) - \bar{\boldsymbol{\mu}}_{o_s} \right] = 0, \quad \text{or} \qquad (5.23)$$

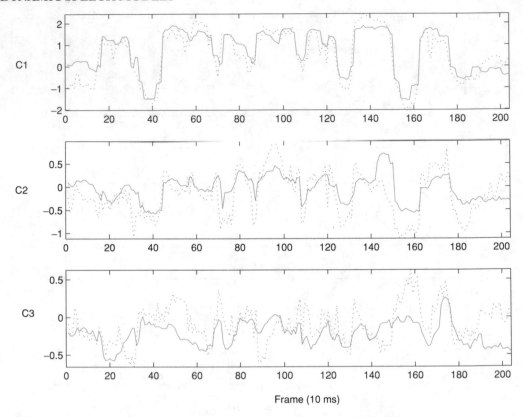

FIGURE 5.11: Same as Fig. 5.10 except with the second utterance (SI2299)

$$\sum_{k=1}^{K}\Big[o(k) - \mathcal{F}'[z_0(k)]\boldsymbol{\mu}_z(k)$$
$$-\{\mathcal{F}[z_0(k)] + \boldsymbol{\mu}_{r_s} - \mathcal{F}'[z_0(k)]z_0(k)\}\Big] = 0. \tag{5.24}$$

Solving for $\boldsymbol{\mu}_{r_s}$, we have the estimation formula of

$$\hat{\boldsymbol{\mu}}_{r_s} = \frac{\sum_k \Big[o(k) - \mathcal{F}[z_0(k)] - \mathcal{F}'[z_0(k)]\boldsymbol{\mu}_z(k) + \mathcal{F}'[z_0(k)]z_0(k)\Big]}{K}. \tag{5.25}$$

Diagonal Covariance Matrices

Denote the diagonal elements of the covariance matrices for the residuals as a vector $\boldsymbol{\sigma}_{r_s}^2$. To derive the ML estimate, we set

$$\frac{\partial \log \prod_{k=1}^{K} p(o(k)\,|\,s)}{\partial \boldsymbol{\sigma}_{r_s}^2} = 0,$$

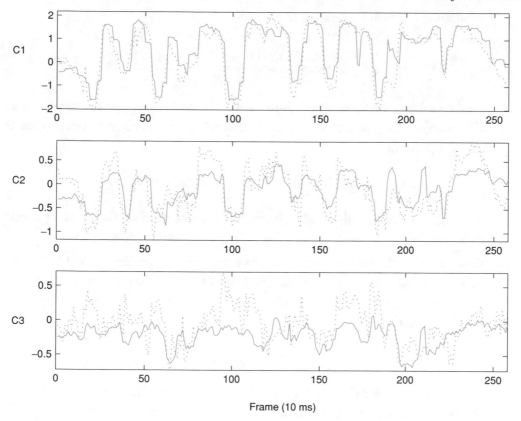

FIGURE 5.12: Same as Fig. 5.10 except with the third utterance (SI1669)

which gives

$$\sum_{k=1}^{K}\left[\frac{\sigma_{r_s}^2 + \mathbf{q}(k) - (\mathbf{o}(k) - \bar{\boldsymbol{\mu}}_{o_s})^2}{[\sigma_{r_s}^2 + \mathbf{q}(k)]^2}\right] = 0, \qquad (5.26)$$

where vector squaring is the element-wise operation, and

$$\mathbf{q}(k) = \text{diag}\Big[\mathcal{F}'[\mathbf{z}_0(k)]\boldsymbol{\Sigma}_z(k)(\mathcal{F}'[\mathbf{z}_0(k)])^{\text{Tr}}\Big]. \qquad (5.27)$$

Due to the frame (k) dependency in the denominator in Eq. (5.26), no simple closed-form solution is available for solving $\sigma_{r_s}^2$ from Eq. (5.26). We have implemented three different techniques for seeking approximate ML estimates that are outlined as follows:

1. *Frame-independent approximation:* Assume the dependency of $\mathbf{q}(k)$ on time frame k is mild, or $\mathbf{q}(k) \approx \bar{\mathbf{q}}$. Then the denominator in Eq. (5.26) can be cancelled, yielding the

approximate closed-form estimate of

$$\hat{\sigma}_{r_s}^2 \approx \frac{\sum_{k=1}^{K} \left\{ (o(k) - \bar{\mu}_{o_s})^2 - \mathbf{q}(k) \right\}}{K}. \qquad (5.28)$$

2. *Direct gradient ascent:* Make no assumption of the above, and take the left-hand side of Eq. (5.26) as the gradient ∇L of log-likelihood of the data in the standard gradient-ascent algorithm:

$$\sigma_{r_s}^2(t+1) = \sigma_{r_s}^2(t) + \epsilon_t \nabla L(o_1^K \mid \sigma_{r_s}^2(t)),$$

where ϵ_t is a heuristically chosen positive constant controlling the learning rate at the t-th iteration.

3. *Constrained gradient ascent:* Add to the previous standard gradient ascent technique the constraint that the variance estimate be always positive. The constraint is established by the parameter transformation: $\tilde{\sigma}_{r_s}^2 = \log \sigma_{r_s}^2$, and by performing gradient ascent for $\tilde{\sigma}_{r_s}^2$ instead for $\sigma_{r_s}^2$:

$$\tilde{\sigma}_{r_s}^2(t+1) = \tilde{\sigma}_{r_s}^2(t) + \tilde{\epsilon}_t \nabla \tilde{L}(o_1^K \mid \tilde{\sigma}_{r_s}^2(t)),$$

Using chain rule, we show below that the new gradient $\nabla \tilde{L}$ is related to the gradient ∇L before parameter transformation in a simple manner:

$$\nabla \tilde{L} = \frac{\partial \tilde{L}}{\partial \tilde{\sigma}_{r_s}^2} = \frac{\partial \tilde{L}}{\partial \sigma_{r_s}^2} \frac{\partial \sigma_{r_s}^2}{\partial \tilde{\sigma}_{r_s}^2} = (\nabla L) \, \exp(\tilde{\sigma}_{r_s}^2).$$

At the end of the algorithm iteration, the parameters are transformed via $\sigma_{r_s}^2 = \exp(\tilde{\sigma}_{r_s}^2)$, which is guaranteed to be positive.

For efficiency purposes, parameter updating in the above gradient ascent techniques is carried out after each utterance in the training, rather than after the entire batch of all utterances.

We note that the quality of the estimates for the residual parameters discussed above plays a crucial role in phonetic recognition performance. These parameters provide an important mechanism for distinguishing speech sounds that belong to different manners of articulation. This is attributed to the fact that nonlinear cepstral prediction from VTRs has different accuracy for these different classes of sounds. Within the same manner class, the phonetic separation is largely accomplished by distinct VTR targets, which typically induce significantly different cepstral prediction values via the "amplification" mechanism provided by the Jacobian matrix $\mathcal{F}'[z]$.

5.3.2   Vocal Tract Resonance Targets' Distributional Parameters

This subset of the HTM parameters consists of (1) the mean vectors $\boldsymbol{\mu}_{T_s}$ and (2) the diagonal elements $\sigma_{T_s}^2$ in the covariance matrices of the stochastic segmental VTR targets. They are also conditioned on phone segment s (and not on sub-phone segment).

Mean Vectors

To obtain a closed-form estimation solution, we assume diagonality of the prediction cepstral residual's covariance matrix $\boldsymbol{\Sigma}_{r_s}$. Denoting its qth component by $\sigma_r^2(q)$ ($q = 1, 2, \ldots, Q$), we decompose the multivariate Gaussian of Eq. (5.20) element-by-element into

$$p(\boldsymbol{o}(k) \mid s(k)) = \prod_{j=1}^{J} \frac{1}{\sqrt{2\pi \sigma_{o_{s(k)}}^2(j)}} \exp\left\{-\frac{(o_k(j) - \bar{\mu}_{o_{s(k)}}(j))^2}{2\sigma_{o_{s(k)}}^2(j)}\right\}, \qquad (5.29)$$

where $o_k(j)$ denotes the jth component (i.e., jth order) of the cepstral observation vector at frame k.

The log-likelihood function for a training data sequence ($k = 1, 2, \ldots, K$) relevant to the VTR mean vector $\boldsymbol{\mu}_{T_s}$ becomes

$$P = \sum_{k=1}^{K} \sum_{j=1}^{J} \left\{-\frac{(o_k(j) - \bar{\mu}_{o_{s(k)}}(j))^2}{\sigma_{o_{s(k)}}^2(j)}\right\} \qquad (5.30)$$

$$= \sum_{k=1}^{K} \sum_{j=1}^{J} \left\{\frac{[\sum_f \mathcal{F}'[z_0(k), j, f] \sum_l a_k(l)\mu_T(l, f) - d_k(j)]^2}{\sigma_{o_{s(k)}}^2(j)}\right\},$$

where l and f are indices to phone and to VTR component, respectively, and

$$d_k(j) = o_k(j) - F[z_0(k), j] + \sum_f \mathcal{F}'[z_0(k), j, f]z_0(k, f) - \mu_{r_{s(k)}}(j).$$

While the acoustic feature's distribution is Gaussian for both HTM and HMM given the state s, the key difference is that the mean and variance in HTM as in Eq. (5.20) are both time-varying functions (hence trajectory model). These functions provide context dependency (and possible target undershooting) via the smoothing of targets across phonetic units in the utterance. This smoothing is explicitly represented in the weighted sum over all phones in the utterance (i.e., $\sum_l$) in Eq. (5.30).

Setting

$$\frac{\partial P}{\partial \mu_T(l_0, f_0)} = 0,$$

and grouping terms involving unknown $\mu_T(l, f)$ on the left and the remaining terms on the right, we obtain

$$
\sum_f \sum_l A(l, f; l_0, f_0) \mu_T(l, f)
$$
$$
= \sum_k \left\{ \sum_j \frac{\mathcal{F}'[z_0(k), j, f_0]}{\sigma^2_{o_{s(k)}}(j)} d_k(j) \right\} a_k(l_0), \tag{5.31}
$$

with $f_0 = 1, 2, \ldots, 8$ for each VTR dimension, and with $l_0 = 1, 2, \ldots, 58$ for each phone unit. In Eq. (5.31),

$$
A(l, f; l_0, f_0) = \sum_{k,j} \frac{\mathcal{F}'[z_0(k), j, f] \mathcal{F}'[z_0(k), j, f_0]}{\sigma^2_{o_{s(k)}}(j)} a_k(l_0) a_k(l). \tag{5.32}
$$

Eq. (5.31) is a 464×464 full-rank linear system of equations. (The dimension $464 = 58 \times 8$ where we have a total of 58 phones in the TIMIT database after decomposing each diphthong into two "phones", and 8 is the VTR vector dimension.) Matrix inversion gives an ML estimate of the complete set of target mean parameters: a 464-dimensional vector formed by concatenating all eight VTR components (four frequencies and four bandwidths) of the 58 phone units in TIMIT.

In implementing Eq. (5.31) for the ML solution to target mean vectors, we kept other model parameters constant. Estimation of the target and residual parameters was carried out in an iterative manner. Initialization of the parameters $\mu_T(l, f)$ was provided by the values described in [9].

An alternative training of the target mean parameters in a simplified version of the HTM and its experimental evaluation are described in [112]. In that training, the VTR tracking results obtained by the tracking algorithm described in Chapter 4 are exploited as the basis for learning, contrasting the learning described in this section, which uses the raw cepstral acoustic data only. Use of the VTR tracking results enables speaker-adaptive learning for the VTR target parameters as shown in [112].

Diagonal Covariance Matrices

To establish the objective function for optimization, we take logarithm on the sum of the likelihood function Eq. (5.29) (over K frames) to obtain

$$
L_T \propto - \sum_{k=1}^{K} \sum_{j=1}^{J} \left\{ \frac{(o_k(j) - \bar{\mu}_{o_{s(k)}}(j))^2}{\sigma^2_{r_s}(j) + q(k, j)} + \log[\sigma^2_{r_s}(j) + q(k, j)] \right\}, \tag{5.33}
$$

where $q(k, j)$ is the jth element of the vector $\mathbf{q}(k)$ as defined in Eq. (5.27). When $\mathbf{\Sigma}_z(k)$ is diagonal, it can be shown that

$$q(k, j) = \sum_f \sigma^2_{z(k)}(f)(F'_{jf})^2 = \sum_f \sum_l v_k(l)\sigma^2_T(l, f)(F'_{jf})^2, \tag{5.34}$$

where F'_{jf} is the (j, f) element of Jacobian matrix $\mathcal{F}'[\cdot]$ in Eq. (5.27), and the second equality is due to Eq. (5.11).

Using chain rule to compute the gradient, we obtain

$$\nabla L_T(l, f) = \frac{\partial L_T}{\partial \sigma^2_T(l, f)} \tag{5.35}$$

$$= \sum_{k=1}^{K} \sum_{j=1}^{J} \left\{ \frac{(o_k(j) - \bar{\mu}_{o_{s(k)}}(j))^2(F'_{jf})^2 v_k(l)}{[\sigma^2_{r_s}(j) + q(k, j)]^2} - \frac{(F'_{jf})^2 v_k(l)}{\sigma^2_{r_s}(j) + q(k, j)} \right\}.$$

Gradient-ascend iterations then proceed as follows:

$$\sigma^2_T(l, f) \leftarrow \sigma^2_T(l, f) + \epsilon \nabla L_T(l, f),$$

for each phone l and for each element f in the diagonal VTR target covariance matrix.

5.4   APPLICATION TO PHONETIC RECOGNITION

5.4.1   Experimental Design

Phonetic recognition experiments have been conducted [124] aimed at evaluating the HTM and the parameter learning algorithms described in this chapter. The standard TIMIT phone set with 48 labels is expanded to 58 (as described in [9]) in training the HTM parameters using the standard training utterances. Phonetic recognition errors are tabulated using the commonly adopted 39 labels after the label folding. The results are reported on the standard core test set of 192 utterances by 24 speakers [127].

Due to the high implementation and computational complexity for the full-fledged HTM decoder, the experiments reported in [124] have been restricted to those obtained by N-best rescoring and lattice-constrained search. For each of the core test utterances, a standard decision-tree-based triphone HMM with the bi-gram language model is used to generate a large N-best list ($N = 1000$) and a large lattice. These N-best lists and lattices are used for the rescoring experiments with the HTM. The HTM system is trained using the parameter estimation algorithms described earlier in this chapter. Learning rates in the gradient ascent techniques have been tuned empirically.

5.4.2   Experimental Results

In Table 5.1, phonetic recognition performance comparisons are shown between the HMM system described above and three evaluation versions of the HTM system. The HTM-1 version uses the HTM likelihood computed from Eq. (5.20) to rescore the 1000-best lists, and no HMM score and language model (LM) score attached in the 1000-best list are exploited. The HTM-2 version improves the HTM-1 version slightly by linearly weighting the log-likelihoods of the HTM, the HMM and the (bigram) LM, based on the same 1000-best lists. The HTM-3 version replaces the 1000-best lists by the lattices, and carries out A* search, constrained by the lattices and with linearly weighted HTM–HMM–LM scores, to decode phonetic sequences. (See a detailed technical description of this A*-based search algorithm in [111].) Notable performance improvement is obtained as shown in the final row of Table 5.1. For all the systems, the performance is measured by percent phone recognition accuracy (i.e., including insertion errors) averaged over the core test-set sentences (numbers in bolds in column 2). The percent-correctness performance (i.e., excluding insertion errors) is listed in column 3. The substitution, deletion and insertion error rates are shown in the remaining columns.

The performance results in Table 5.1 are obtained using the identical acoustic features of frequency-warped linear cepstra for all the systems. Frequency warping of linear cepstra [128] has been implemented by a linear matrix-multiplication technique on both acoustic features and the observation-prediction component of the HTM. The warping gives slight performance improvement for both HMM and HTM systems by a similar amount. Overall, the lattice-based HTM system (75.07% accuracy) gives 13% fewer errors than does the HMM system

TABLE 5.1: TIMIT Phonetic Recognition Performance Comparisons Between an HMM System and Three Versions of the HTM System

	ACC %	CORR %	SUB %	DEL %	INS %
HMM	71.43	73.64	17.14	9.22	2.21
HTM-1	74.31	77.76	16.23	6.01	3.45
HTM-2	74.59	77.73	15.61	6.65	3.14
HTM-3	75.07	78.28	15.94	5.78	3.20

Note. HTM-1: *N*-best rescoring with HTM scores only; HTM-2: *N*-best rescoring with weighted HTM, HMM and LM scores; HTM-3: Lattice-constrained A* search with weighted HTM, HMM and LM scores. Identical acoustic features (frequency-warped linear cepstra) are used.

(71.43% accuracy). This performance is better than that of any HMM system on the same task as summarized in [127], and is approaching the best-ever result (75.6% accuracy) obtained by using many heterogeneous classifiers as reported in [127] also.

5.5   SUMMARY

In this chapter, we present in detail a second specific type of hidden dynamic models, which we call the hidden trajectory model (HTM). The unique character of the HTM is that the hidden dynamics are represented not by temporal recursion on themselves but by explicit "trajectories" or hidden trended functions constructed by FIR filtering of targets. In contrast to the implementation strategy for the model discussed in Chapter 4 where the hidden dynamics are discretized, the implementation strategy in the HTM maintains continuous-valued hidden dynamics, and introduces approximations by constraining the temporal boundaries associated with discrete phonological states. Given such constraints, rigorous algorithms for model parameter estimation are developed and presented without the need to approximate the continuous hidden dynamic variables by their discretized values as done in Chapter 4.

The main portions of this chapter are devoted to formal construction of the HTM, its computer simulation and the parameter estimation algorithm's development. The computationally efficient decoding algorithms have not been presented, as they are still under research and development and are hence not appropriate to describe in this book at present. In contrast, decoding algorithms for discretized hidden dynamic models are much more straightforward to develop, as we have presented in Chapter 4.

Although we present only two types of implementation strategies in this book (Chapters 4, 5, respectively) for dynamic speech modeling within the general computational framework established in Chapter 2, other types of implementation strategies and approximations (such as variational learning and decoding) are possible. We have given some related references at the beginning of this chapter.

As a summary and conclusion of this book, we have provided scientific background, mathematical theory, computational framework, algorithmic development and technological needs and two selected applications for dynamic speech modeling, which is the theme of this book. A comprehensive survey in this area of research is presented, drawing on the work of a number of (non-exhaustive) research groups and individual researchers worldwide. This direction of research is guided by scientific principles applied to study human speech communication, and is based on the desire to acquire knowledge about the realistic dynamic process in the closed-loop speech chain. It is hoped that with integration of this unique style of research with other powerful pattern recognition and machine learning approaches, the dynamic speech models, as they become better developed, will form a foundation for the next-generation speech technology serving the humankind and society.

Bibliography

[1] P. Denes and E. Pinson. *The Speech Chain*, 2nd edn, Worth Publishers, New York, 1993.

[2] K. Stevens. Acoustic Phonetics, MIT Press, Cambridge, MA, 1998.

[3] K. Stevens. "Toward a model for lexical access based on acoustic landmarks and distinctive features," *J. Acoust. Soc. Am.*, Vol. 111, April 2002, pp. 1872–1891. doi:10.1121/1.1458026

[4] L. Rabiner and B.-H. Juang. *Fundamentals of Speech Recognition*, Prentice-Hall, Upper Saddle River, NJ, 1993.

[5] X. Huang, A. Acero, and H. Hon. *Spoken Language Processing*, Prentice Hall, New York, 2001.

[6] V. Zue. "Notes on speech spectrogram reading," MIT Lecture Notes, Cambridge, MA, 1991.

[7] J. Olive, A. Greenwood, and J. Coleman. *Acoustics of American English Speech—A Dynamic Approach*, Springer-Verlag, New York, 1993.

[8] C. Williams. "How to pretend that correlated variables are independent by using difference observations," *Neural Comput.*, Vol. 17, 2005, pp. 1–6. doi:10.1162/0899766052530884

[9] L. Deng and D. O'Shaughnessy. *SPEECH PROCESSING—A Dynamic and Optimization-Oriented Approach* (ISBN: 0-8247-4040-8), Marcel Dekker, New York, 2003, pp. 626.

[10] L. Deng and X.D. Huang. "Challenges in adopting speech recognition," *Commun. ACM*, Vol. 47, No. 1, January 2004, pp. 69–75.

[11] M. Ostendorf. "Moving beyond the beads-on-a-string model of speech," in *Proceedings of IEEE Workshop on Automatic Speech Recognition and Understanding*, December 1999, Keystone, co, pp. 79–83.

[12] N. Morgan, Q. Zhu, A. Stolcke, *et al.* "Pushing the envelope—Aside," *IEEE Signal Process. Mag.*, Vol. 22, No. 5, September. 2005, pp. 81–88.doi:10.1109/MSP.2005.1511826

[13] F. Pereira. "Linear models for structure prediction," in *Proceedings of Interspeech*, Lisbon, September 2005, pp. 717–720.

[14] M. Ostendorf, V. Digalakis, and J. Rohlicek. "From HMMs to segment models: A unified view of stochastic modeling for speech recognition" *IEEE Trans. Speech Audio Process.*, Vol. 4, 1996, pp. 360–378.doi:10.1109/89.536930

[15] B.-H. Juang and S. Katagiri. "Discriminative learning for minimum error classification," *IEEE Trans. Signal Process.*, Vol. 40, No. 12, 1992, pp. 3043–3054. doi:10.1109/78.175747

[16] D. Povey. "Discriminative training for large vocabulary speech recognition," Ph.D. dissertation, Cambridge University, 2003.

[17] W. Chou and B.-H. Juang (eds.). *Pattern Recognition in Speech and Language Processing,* CRC Press, Boca Raton, FL, 2003.

[18] L. Deng, J. Wu, J. Droppo, and A. Acero. "Analysis and comparison of two feature extraction/compensation algorithms," *IEEE Signal Process. Lett.*, Vol. 12, No. 6, June 2005, pp. 477–480.doi:10.1109/LSP.2005.847861

[19] D. Povey, B. Kingsbury, L. Mangu, G. Saon, H. Solatu, and G. Zweig. "FMPE: Discriminatively trained features for speech recognition," *IEEE Proc. ICASSP*, Vol. 2, 2005, pp. 961–964.

[20] J. Bilmes and C. Bartels. "Graphical model architectures for speech recognition," *IEEE Signal Process. Mag.*, Vol. 22, No. 5, Sept. 2005, pp. 89–100. doi:10.1109/MSP.2005.1511827

[21] G. Zweig. "Bayesian network structures and inference techniques for automatic speech recognition," *Comput. Speech Language*, Vol. 17, No. 2/3, 2003, pp. 173–193.

[22] F. Jelinek, *et al.* "Central issues in the recognition of conversational speech," Summary Report, Johns Hopkins University, Baltimore, MD, 1999, pp. 1–57.

[23] S. Greenberg, J. Hollenback, and D. Ellis. "Insights into spoken language gleaned from phonetic transcription of the Switchboard corpus," *Proc. ICSLP*, Vol. 1, 1996, pp. S32–S35.

[24] L. Deng and J. Ma. "Spontaneous speech recognition using a statistical coarticulatory model for the hidden vocal–tract–resonance dynamics," *J. Acoust. Soc. Am.*, Vol. 108, No. 6, 2000, pp. 3036–3048.doi:10.1121/1.1315288

[25] S. Furui, K. Iwano, C. Hori, T. Shinozaki, Y. Saito, and S. Tamur. "Ubiquitous speech processing," *IEEE Proc. ICASSP*, Vol. 1, 2001, pp. 13–16.

[26] K.C. Sim and M. Gales. "Temporally varying model parameters for large vocabulary continuous speech recognition," in *Proceedings of Interspeech*, Lisbon, September 2005, pp. 2137–2140.

[27] K.-F. Lee. *Automatic speech recognition: The Development of the Sphinx Recognition System,* Springer, New York, 1988.

[28] C.-H. Lee, F. Soong, and K. Paliwal (eds.). *Automatic Speech and Speaker Recognition— Advanced Topics,* Kluwer Academic, Norwell, MA, 1996.

[29] F. Jelinek. *Statistical Methods for Speech Recognition,* MIT Press, Cambridge, MA, 1997.

[30] B.-H. Juang and S. Furui (Eds.). *Proc. IEEE* (special issue), Vol. 88, 2000.

[31] L. Deng, K. Wang, and W. Chou. "Speech technology and systems in human–Machine communication—Guest editors' editorial," *IEEE Signal Process. Mag.*, Vol. 22, No. 5, September 2005, pp. 12–14.doi:10.1109/MSP.2005.1511818

[32] J. Allen. "How do humans process and recognize speech," *IEEE Trans. Speech Audio Process.*, Vol. 2, 1994, pp. 567–577.doi:10.1109/89.326615

[33] L. Deng. "A dynamic, feature-based approach to the interface between phonology and phonetics for speech modeling and recognition," *Speech Commun.*, Vol. 24, No. 4, 1998, pp. 299–323.doi:10.1016/S0167-6393(98)00023-5

[34] H. Bourlard, H. Hermansky, and N. Morgan. "Towards increasing speech recognition error rates," *Speech Commun.*, Vol. 18, 1996, pp. 205–231.doi:10.1016/0167-6393(96)00003-9

[35] L. Deng. "Switching dynamic system models for speech articulation and acoustics," in M. Johnson, M. Ostendorf, S. Khudanpur, and R. Rosenfeld (eds.), *Mathematical Foundations of Speech and Language Processing*, Springer-Verlag, New York, 2004, pp. 115–134.

[36] R. Lippmann. "Speech recognition by human and machines," *Speech Commun.*, Vol. 22, 1997, pp. 1–14.doi:10.1016/S0167-6393(97)00021-6

[37] L. Pols. "Flexible human speech recognition," in *Proceedings of the IEEE Workshop on Automatic Speech Recognition and Understanding*, 1997, Santa Barbara, CA, pp. 273–283.

[38] C.-H. Lee. "From knowledge-ignorant to knowledge-rich modeling: A new speech research paradigm for next-generation automatic speech recognition," in *Proc. ICSLP*, Jeju Island, Korea, October 2004, pp. 109–111.

[39] M. Russell. "Progress towards speech models that model speech," in *Proc. IEEE Workshop on Automatic Speech Recognition and Understanding*, 1997, Santa Barbara, CA, pp. 115–123.

[40] M. Russell. "A segmental HMM for speech pattern matching," *IEEE Proceedings of the ICASSP*, Vol. 1, 1993, pp. 499–502.

[41] L. Deng. "A generalized hidden Markov model with state-conditioned trend functions of time for the speech signal," *Signal Process.*, Vol. 27, 1992, pp. 65–78.doi:10.1016/0165-1684(92)90112-A

[42] J. Bridle, L. Deng, J. Picone, *et al.* "An investigation of segmental hidden dynamic models of speech coarticulation for automatic speech recognition," Final Report for the 1998 Workshop on Language Engineering, Center for Language and Speech Processing at Johns Hopkins University, 1998, pp. 1–61.

[43] K. Kirchhoff. "Robust speech recognition using articulatory information," Ph.D. thesis, University of Bielfeld, Germany, July 1999.

[44] R. Bakis. "Coarticulation modeling with continuous-state HMMs," in *Proceedings of the IEEE Workshop on Automatic Speech Recognition*, Harriman, New York, 1991, pp. 20–21.

[45] Y. Gao, R. Bakis, J. Huang, and B. Zhang. "Multistage coarticulation model combining articulatory, formant and cepstral features," *Proc. ICSLP*, Vol. 1, 2000, pp. 25–28.

[46] J. Frankel and S. King. "ASR—Articulatory speech recognition," *Proc. Eurospeech*, Vol. 1, 2001, pp 599–602.

[47] T. Kaburagi and M. Honda. "Dynamic articulatory model based on multidimensional invariant-feature task representation," *J. Acoust. Soc. Am.*, 2001, Vol. 110, No. 1, pp. 441–452.

[48] P. Jackson, B. Lo, and M. Russell. "Data-driven, non-linear, formant-to-acoustic mapping for ASR," *IEE Electron. Lett.*, Vol. 38, No. 13, 2002, pp. 667–669. doi:10.1049/el:20020436

[49] M. Russell and P. Jackson. "A multiple-level linear/linear segmental HMM with a formant-based intermediate layer," *Comput. Speech Language*, Vol. 19, No. 2, 2005, pp. 205–225.doi:10.1016/j.csl.2004.08.001

[50] L. Deng and D. Sun. "A statistical approach to automatic speech recognition using the atomic speech units constructed from overlapping articulatory features," *J. Acoust. Soc. Am.*, Vol. 95, 1994, pp. 2702–2719.doi:10.1121/1.409839

[51] H. Nock and S. Young. "Loosely coupled HMMs for ASR: A preliminary study," Technical Report TR386, Cambridge University, 2000.

[52] K. Livescue, J. Glass, and J. Bilmes. "Hidden feature models for speech recognition using dynamic Bayesian networks," *Proc. Eurospeech*, Vol. 4, 2003, pp. 2529–2532.

[53] E. Saltzman and K. Munhall. "A dynamical approach to gestural patterning in speech production," *Ecol. Psychol.*, Vol. 1, pp. 333–382.

[54] L. Deng. "Computational models for speech production," in K. Ponting (ed.), *Computational Models of Speech Pattern Processing (NATO ASI Series)*, Springer, New York, 1999, pp. 199–214.

[55] L. Deng, M. Aksmanovic, D. Sun, and J. Wu. "Speech recognition using hidden Markov models with polynomial regression functions as nonstationary states," *IEEE Trans. Speech Audio Process.*, Vol. 2, 1994, pp. 507–520.doi:10.1109/89.326610

[56] C. Li and M. Siu, "An efficient incremental likelihood evaluation for polynomial trajectory model with application to model training and recognition," *IEEE Proc. ICASSP*, Vol. 1, 2003, pp. 756–759.

[57] Y. Minami, E. McDermott, A. Nakamura, and S. Katagiri. "Recognition method with parametric trajectory generated from mixture distribution HMMs," *IEEE Proc. ICASSP*, Vol. 1, 2003, pp. 124–127.

[58] C. Blackburn and S. Young. "A self-learning predictive model of articulator movements during speech production," *J. Acoust. Soc. Am.*, Vol. 107, No. 3, 2000, pp. 1659–1670. doi:10.1121/1.428450

[59] L. Deng, G. Ramsay, and D. Sun. "Production models as a structural basis for automatic speech recognition," *Speech Commun.*, Vol. 22, No. 2, 1997, pp. 93–111. doi:10.1016/S0167-6393(97)00018-6

[60] B. Lindblom. "Explaining phonetic variation: A sketch of the H & H theory," in W. Hardcastle and A. Marchal (eds.), *Speech Production and Speech Modeling*, Kluwer, Norwell, MA, 1990, pp. 403–439.

[61] N. Chomsky and M. Halle. *The Sound Pattern of English*, Harper and Row, New York, 1968.

[62] N. Clements. "The geometry of phonological features," *Phonology Yearbook*, Vol. 2, 1985, pp. 225–252.

[63] C. Browman and L. Goldstein. "Articulatory phonology: An overview," *Phonetica*, Vol. 49, 1992, pp. 155–180.

[64] M. Randolph. "Speech analysis based on articulatory behavior," *J. Acoust. Soc. Am.*, Vol. 95, 1994, p. 195.

[65] L. Deng and H. Sameti. "Transitional speech units and their representation by the regressive Markov states: Applications to speech recognition," *IEEE Trans. Speech Audio Process.*, Vol. 4, No. 4, July 1996, pp. 301–306.doi:10.1109/89.506934

[66] J. Sun, L. Deng, and X. Jing. "Data-driven model construction for continuous speech recognition using overlapping articulatory features," *Proc. ICSLP*, Vol. 1, 2000, pp. 437–440.

[67] Z. Ghahramani and M. Jordan. "Factorial hidden Markov models," *Machine Learn.*, Vol. 29, 1997, pp.245–273.doi:10.1023/A:1007425814087

[68] K. Stevens. "On the quantal nature of speech," *J. Phonetics*, Vol. 17, 1989, pp. 3–45.

[69] A. Liberman and I. Mattingly. "The motor theory of speech perception revised," *Cognition*, Vol. 21, 1985, pp. 1–36.doi:10.1016/0010-0277(85)90021-6

[70] B. Lindblom. "Role of articulation in speech perception: Clues from production," *J. Acoust. Soc. Am.*, Vol. 99, No. 3, 1996, pp. 1683–1692. doi:10.1121/1.414691

[71] P. MacNeilage. "Motor control of serial ordering in speech," *Psychol. Rev.*, Vol. 77, 1970, pp. 182–196.

[72] R. Kent, G. Adams, and G. Turner. "Models of speech production," in N. Lass (ed.), *Principles of Experimental Phonetics*, Mosby, London, 1995, pp. 3–45.

[73] J. Perkell, M. Matthies, M. Svirsky, and M. Jordan. "Goal-based speech motor control: A theoretical framework and some preliminary data," *J. Phonetics*, Vol. 23, 1995, pp. 23–35.doi:10.1016/S0095-4470(95)80030-1

[74] J. Perkell. "Properties of the tongue help to define vowel categories: Hypotheses based on physiologically-oriented modeling," *J. Phonetics*, Vol. 24, 1996, pp. 3–22. doi:10.1006/jpho.1996.0002

[75] P. Perrier, D. Ostry, and R. Laboissière. "The equilibrium point hypothesis and its application to speech motor control," *J. Speech Hearing Res.*, Vol. 39, 1996, pp. 365–378.

[76] B. Lindblom, J. Lubker, and T. Gay. "Formant frequencies of some fixed-mandible vowels and a model of speech motor programming by predictive simulation," *J. Phonetics*, Vol. 7, 1979, pp. 146–161.

[77] S. Maeda. "On articulatory and acoustic variabilities," *J. Phonetics*, Vol. 19, 1991, pp. 321–331.

[78] G. Ramsay and L. Deng. "A stochastic framework for articulatory speech recognition," *J. Acoust. Soc. Am.*, Vol. 95, No. 6, 1994, p. 2871.

[79] C. Coker. "A model of articulatory dynamics and control," *Proc. IEEE*, Vol. 64, No. 4, 1976, pp. 452–460.

[80] P. Mermelstein. "Articulatory model for the study of speech production," *J. Acoust. Soc. Am.*, Vol. 53, 1973, pp. 1070–1082.doi:10.1121/1.1913427

[81] C. Bishop. *Neural Networks for Pattern Recognition*, Clarendon Press, Oxford, 1995.

[82] Z. Ghahramani and S. Roweis. "Learning nonlinear dynamic systems using an EM algorithm," *Adv. Neural Informat. Process. Syst.*, Vol. 11, 1999, pp. 1–7.

[83] L. Deng, J. Droppo, and A. Acero. "Estimating cepstrum of speech under the presence of noise using a joint prior of static and dynamic features," *IEEE Trans. Speech Audio Process.*, Vol. 12, No. 3, May 2004, pp. 218–233.doi:10.1109/TSA.2003.822627

[84] J. Ma and L. Deng. "Target-directed mixture linear dynamic models for spontaneous speech recognition," *IEEE Trans. Speech Audio Process.*, Vol. 12, No. 1, 2004, pp. 47–58. doi:10.1109/TSA.2003.818074

[85] J. Ma and L. Deng. "A mixed-level switching dynamic system for continuous speech recognition," *Comput. Speech Language*, Vol. 18, 2004, pp. 49–65.doi:10.1016/S0885-2308(03)00031-7

[86] H. Gish and K. Ng. "A segmental speech model with applications to word spotting," *IEEE Proc. ICASSP*, Vol. 1, 1993, pp. 447–450.

[87] L. Deng and M. Aksmanovic. "Speaker-independent phonetic classification using hidden Markov models with mixtures of trend functions," *IEEE Trans. Speech Audio Process.*, Vol. 5, 1997, pp. 319–324.doi:10.1109/89.593305

[88] H. Hon and K. Wang. "Unified frame and segment based models for automatic speech recognition," *IEEE Proc. the ICASSP*, Vol. 2, 2000, pp. 1017–1020.

[89] M. Gales and S. Young. "Segmental HMMs for speech recognition," *Proc. Eurospeech*, Vol. 3, 1993, pp. 1579–1582.

[90] W. Holmes and M. Russell. "Probabilistic-trajectory segmental HMMs," *Comput. Speech Language*, Vol. 13, 1999, pp. 3–27.doi:10.1006/csla.1998.0048

[91] C. Rathinavelu and L. Deng. "A maximum a posteriori approach to speaker adaptation using the trended hidden Markov model," *IEEE Trans. Speech Audio Process.*, Vol. 9, 2001, pp. 549–557.doi:10.1109/89.928919

[92] O. Ghitza and M. Sondhi. "Hidden Markov models with templates as nonstationary states: An application to speech recognition," *Comput. Speech Language*, Vol. 7, 1993, pp. 101–119.doi:10.1006/csla.1993.1005

[93] P. Kenny, M. Lennig, and P. Mermelstein. "A linear predictive HMM for vector-valued observations with applications to speech recognition," *IEEE Trans. Acoust., Speech, Signal Process.*, Vol. 38, 1990, pp. 220–225.doi:10.1109/29.103057

[94] L. Deng and C. Rathinavalu. "A Markov model containing state-conditioned second-order nonstationarity: Application to speech recognition," *Comput. Speech Language*, Vol. 9, 1995, pp. 63–86.doi:10.1006/csla.1995.0004

[95] A. Poritz. "Hidden Markov models: A guided tour," *IEEE Proc. ICASSP*, Vol. 1, 1988, pp. 7–13.

[96] H. Sheikhazed and L. Deng. "Waveform-based speech recognition using hidden filter models: Parameter selection and sensitivity to power normalization," *IEEE Trans. Speech Audio Process.*, Vol. 2, 1994, pp. 80–91.doi:10.1109/89.260337

[97] H. Zen, K. Tokuda, and T. Kitamura. "A Viterbi algorithm for a trajectory model derived from HMM with explicit relationship between static and dynamic features," *IEEE Proc. ICASSP*, 2004, pp. 837–840.

[98] K. Tokuda, H. Zen, and T. Kitamura. "Trajectory modeling based on HMMs with the explicit relationship between static and dynamic features," *Proc. Eurospeech*, Vol. 2, 2003, pp. 865–868.

[99] J. Tebelskis and A. Waibel. "Large vocabulary recognition using linked predictive neural networks," *IEEE Proc. ICASSP*, Vol. 1, 1990, pp. 437–440.

[100] E. Levin. "Word recognition using hidden control neural architecture," *IEEE Proc. ICASSP*, Vol. 1, 1990, pp. 433–436.

[101] L. Deng, K. Hassanein, and M. Elmasry. "Analysis of correlation structure for a neural predictive model with application to speech recognition," *Neural Networks*, Vol. 7, No. 2, 1994, pp. 331–339.doi:10.1016/0893-6080(94)90027-2

[102] V. Digalakis, J. Rohlicek, and M. Ostendorf. "ML estimation of a stochastic linear system with the *EM* algorithm and its application to speech recognition," *IEEE Trans. Speech Audio Process.*, Vol. 1, 1993, pp. 431–442.doi:10.1109/89.242489

[103] L. Deng. "Articulatory features and associated production models in statistical speech recognition," in K. Ponting (ed.), *Computational Models of Speech Pattern Processing (NATO ASI Series)*, Springer, New York, 1999, pp. 214–224.

[104] L. Lee, P. Fieguth, and L. Deng. "A functional articulatory dynamic model for speech production," *IEEE Proc. ICASSP*, Vol. 2, 2001, pp. 797–800.

[105] R. McGowan. "Recovering articulatory movement from formant frequency trajectories using task dynamics and a genetic algorithm: Preliminary model tests," *Speech Commun.*, Vol. 14, 1994, pp. 19–48.doi:10.1016/0167-6393(94)90055-8

[106] R. McGowan and A. Faber. "Speech production parameters for automatic speech recognition," *J. Acoust. Soc. Am.*, Vol. 101, 1997, p. 28.doi:10.1121/1.418310

[107] J. Picone, S. Pike, R. Reagan, T. Kamm, J. Bridle, L. Deng, Z. Ma, H. Richards, and M. Schuster. "Initial evaluation of hidden dynamic models on conversational speech," *IEEE Proc. ICASSP*, Vol. 1, 1999, pp 109–112.

[108] R. Togneri and L. Deng. "Joint state and parameter estimation for a target-directed nonlinear dynamic system model," *IEEE Trans. Signal Process.*, Vol. 51, No. 12, December 2003, pp. 3061–3070.doi:10.1109/TSP.2003.819013

[109] L. Deng, D. Yu, and A. Acero. "A bi-directional target-filtering model of speech coarticulation and reduction: Two-stage implementation for phonetic recognition," *IEEE Trans. Speech Audio Process.*, Vol. 14, No. 1, Jan. 2006, pp. 256–265. doi:10.1109/TSA.2005.854107

[110] L. Deng, A. Acero, and I. Bazzi. "Tracking vocal tract resonances using a quantized nonlinear function embedded in a temporal constraint," *IEEE Trans. Speech Audio Process.*, Vol. 14, No. 2, March 2006, pp. 425–434.doi:10.1109/TSA.2005.855841

[111] D. Yu, L. Deng, and A. Acero. "Evaluation of a long-contextual-span trajectory model and phonetic recognizer using A* lattice search," in *Proceedings of Interspeech*, Lisbon, September 2005, Vol. 1, pp. 553–556.

[112] D. Yu, L. Deng, and A. Acero. "Speaker-adaptive learning of resonance targets in a hidden trajectory model of speech coarticulation," *Comput. Speech Language*, 2006.

[113] H.B. Richards, and J.S. Bridle. "The HDM: A segmental hidden dynamic model of coarticulation," *IEEE Proc. ICASSP*, Vol. 1, 1999, pp. 357–360.

[114] F. Seide, J. Zhou, and L. Deng. "Coarticulation modeling by embedding a target-directed hidden trajectory model into HMM—MAP decoding and evaluation," *IEEE Proc. ICASSP*, Vol. 2, 2003, pp. 748–751.

[115] L. Deng, X. Li, D. Yu, and A. Acero. "A hidden trajectory model with bi-directional target-filtering: Cascaded vs. integrated implementation for phonetic recognition," *IEEE Proceedings of the ICASSP*, Philadelphia, 2005, pp. 337–340.

[116] L. Deng, D. Yu, and A. Acero. "Learning statistically characterized resonance targets in a hidden trajectory model of speech coarticulation and reduction," *Proceedings of the Eurospeech*, Lisbon, 2005, pp. 1097–1100.

[117]   L. Deng, I. Bazzi, and A. Acero. "Tracking vocal tract resonances using an analytical nonlinear predictor and a target-guided temporal constraint," *Proceedings of the Eurospeech*, Vol. I, Geneva, Switzerland, September 2003, pp. 73–76.

[118]   R. Togneri and L. Deng. "A state–space model with neural-network prediction for recovering vocal tract resonances in fluent speech from Mel-cepstral coefficients," *Comput. Speech Language,* 2006.

[119]   A. Acero. "Formant analysis and synthesis using hidden Markov models," in *Proceedings of the Eurospeech*, Budapest, September 1999.

[120]   C. Huang and H. Wang. "Bandwidth-adjusted LPC analysis for robust speech recognition," *Pattern Recognit. Lett.*, Vol. 24, 2003, pp. 1583–1587.doi:10.1016/S0167-8655(02)00397-5

[121]   L. Lee, H. Attias, and L. Deng. "Variational inference and learning for segmental switching state space models of hidden speech dynamics," in *IEEE Proceedings of the ICASSP*, Vol. I, Hong Kong, April 2003, pp. 920–923.

[122]   L. Lee, L. Deng, and H. Attias. "A multimodal variational approach to learning and inference in switching state space models," in *IEEE Proceedings of the ICASSP*, Montreal, Canada, May 2004, Vol. I, pp. 505–508.

[123]   J. Ma and L. Deng. "Effcient decoding strategies for conversational speech recognition using a constrained nonlinear state–space model for vocal–tract–resonance dynamics," *IEEE Trans. Speech Audio Process.*, Vol. 11, No. 6, 2003, pp. 590–602. doi:10.1109/TSA.2003.818075

[124]   L. Deng, D. Yu, and A. Acero. "A long-contextual-span model of resonance dynamics for speech recognition: Parameter learning and recognizer evaluation," *Proceedings of the IEEE Workshop on Automatic Speech Recognition and Understanding*, Puerto Rico, Nov. 27 – Dec 1, 2005, pp. 1–6 (CDROM).

[125]   M. Pitermann. "Effect of speaking rate and contrastive stress on formant dynamics and vowel perception," *J. Acoust. Soc. Am.*, Vol. 107, 2000, pp. 3425–3437. doi:10.1121/1.429413

[126]   L. Deng, L. Lee, H. Attias, and A. Acero. "A structured speech model with continuous hidden dynamics and prediction-residual training for tracking vocal tract resonances," *IEEE Proceedings of the ICASSP*, Montreal, Canada, 2004, pp. 557–560.

[127]   J. Glass. "A probabilistic framework for segment-based speech recognition," *Comput. Speech Language,* Vol. 17, No. 2/3, pp. 137–152.

[128]   A. Oppenheim and D. Johnson. "Discrete representation of signals," *Proc. IEEE*, Vol. 60, No. 6, 1972, pp. 681–691.

About the Author

Li Deng received the B.Sc. degree in 1982 from the University of Science and Technology of China, Hefei, M.Sc. in 1984 and Ph.D. degree in 1986 from the University of Wisconsin – Madison.

Currently, he is a Principal Researcher at Microsoft Research, Redmond, Washington, and an Affiliate Professor of Electrical Engineering at the University of Washington, Seattle (since 1999). Previously, he worked at INRS-Telecommunications, Montreal, Canada (1986–1989), and served as a tenured Professor of Electrical and Computer Engineering at University of Waterloo, Ontario, Canada (1989–1999), where he taught a wide range of electrical engineering courses including signal and speech processing, digital and analog communications, numerical methods, probability theory and statistics. He conducted sabbatical research at Laboratory for Computer Science at Massachusetts Institute of Technology (1992–1993) and at ATR Interpreting Telecommunications Research Laboratories, Kyoto, Japan (1997–1998). He has published over 200 technical papers and book chapters, and is inventor and co-inventor of numerous U.S. and international patents. He co-authored the book "Speech Processing—A Dynamic and Optimization-Oriented Approach" (2003, Marcel Dekker Publishers, New York), and has given keynotes, tutorials and other invited lectures worldwide. He served on Education Committee and Speech Processing Technical Committee of the IEEE Signal Processing Society (1996–2000), and was Associate Editor for IEEE Transactions on Speech and Audio Processing (2002–2005). He currently serves on Multimedia Signal Processing Technical Committee, and on the editorial boards of IEEE Signal Processing Magazine and of EURASIP Journal on Audio, Speech, and Music Processing. He is a Technical Chair of IEEE International Conference on Acoustics, Speech, and Signal Processing (ICASSP 2004) and General Chair of IEEE Workshop on Multimedia Signal Processing (MMSP 2006). He is Fellow of the Acoustical Society of America and Fellow of the IEEE.

Printed in the United States
by Baker & Taylor Publisher Services